Refrigeration on Fishing Vessels

Refrigeration on Fishing Vessels

J. H. Merritt

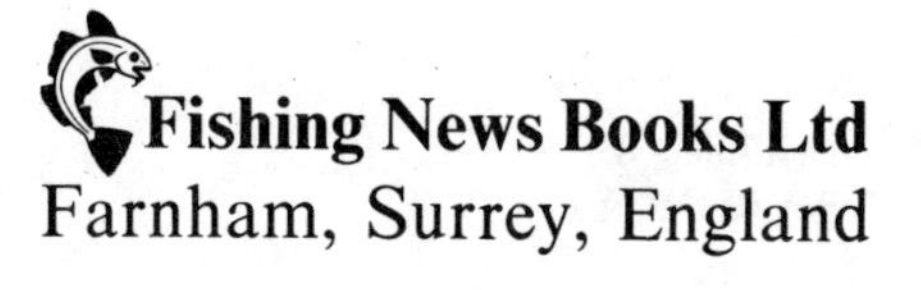

Fishing News Books Ltd
Farnham, Surrey, England

First published 1969
Revised edition 1978

British Library CIP Data

Merritt, John Hugh
Refrigeration on fishing vessels. – Revised ed.
1. Fishing boats – Refrigeration
I. Title
623. 85′35 SH 344. 8. R4

ISBN 0 85238 095 X

Printed in Great Britain by Billing & Sons Limited
Guildford, London and Worcester

Contents

List of Illustrations

Fish Names

Common names for fish have been used in the text. Some scientific names are given below.

anchovy	*Engraulis* sp.
blue whiting	*Micromesistius poutassou*
cod	*Gadus morhua*
crab	all species of the section *Brachyura*
	all species of the family *Lithodidae*
dogfish	*Squalus acanthias*
haddock	*Melanogrammus aeglefinus*
hake	*Merluccius* sp.
halibut	*Hippoglossus hippoglossus*
herring	*Clupea harengus*
lemon sole	*Microstomus kitt*
lobster	*Homarus* sp.
mackerel	*Scomber* sp.
menhaden	*Brevoortia* sp.
Norway lobster	*Nephrops norvegicus*
pilchard	*Sardina pilchardus*
plaice	*Pleuronectes platessa*
pollack	*Pollachius* sp.
prawn	large fish of *Palaemonidae* sp., *Penaeidae* sp., *Pandalidae* sp.
redfish	*Sebastes* sp.
	Helicolenus dactylopterus
saithe	*Pollachius virens*
salmon	*Salmo salar*
	Oncorhynchus sp.
sardine	small *Sardina pilchardus*
shrimp	*Crangon* sp.
	small fish of *Palaemonidae* sp., *Penaeidae* sp., *Pandalidae* sp.
skate	*Raja* sp.
sole	*Solea solea*

sprat	*Sprattus sprattus*
squid	*Loligo* sp.
tuna	*Thunnus* sp. except *Thunnus alalunga, Neothunnus* sp.
turbot	*Scophthalmus maximus*
whiting	*Merlangus merlangus*

The term 'roundfish', used in the text, refers to fish such as cod and hake, as opposed to flatfish such as plaice.

Units

The units used in the text are according to the International System of Units, SI units. A table for conversion of some British units and some non-SI metric units to approximate equivalent values in SI units is given below.

Length		
1 ft	=	0·304 8 m
1 in	=	2·540 cm = 25·40 mm
Area		
1 ft^2	=	0·092 90 m^2 = 9·290 dm^2
1 in^2	=	6·452 cm^2
Volume		
1 ft^3	=	0·028 32 m^3 = 28·32 dm^3
1 in^3	=	16·39 cm^3
Capacity		
1 gal	=	4·546 dm^3 = 4·546 litres
1 USgal	=	3·785 dm^3 = 3·785 litres
Velocity		
1 ft/s	=	0·304 8 m/s = 30·48 cm/s
Mass		
1 t (metric ton)	=	1000 kg
1 ton	=	1016 kg
1 cwt	=	50·80 kg
1 stone	=	6·350 kg
1 lb	=	0·453 6 kg
Mass per unit area		
1 lb/ft^2	=	4·882 kg/m^2
Specific volume		
1 ft^3/lb	=	62·43 dm^3/kg
Mass rate of flow		
1 lb/s	=	0·453 6 kg/s
Volume rate of flow		
1 ft^3/s	=	0·028 32 m^3/s = 28·32 dm^3/s
1 gal/s	=	4·546 dm^3/s

Density

1 lb/ft^3	=	16·02 kg/m^3
Force		
1 kgf or kp	=	9·807 N ($kg\ m/s^2$)
1 lbf (pound-force)	=	4·448 N ($kg\ m/s^2$)
Pressure		
1 atm	=	101·3 kN/m^2
1 b (bar)	=	100·0 kN/m^2
1 kgf/m^2 (mmH_2O)	=	9·807 N/m^2
1 lbf/in^2	=	6895 N/m^2 = 68·95 mb
Energy		
1 cal	=	4·187 J
1 kW h	=	3·600 MJ
1 Btu	=	1·055 kJ
Power		
1 horse power	=	745·7 W (J/s) = 0·745 7 kW
Temperature		
5(°F − 32)/9	=	°C
Heat flow rate		
1 cal/s	=	4·187 J/s (W)
1 kcal/h	=	1·163 J/s (W)
1 Btu/h	=	0·293 1 J/s (W)
Specific energy		
1 kcal/kg	=	4·187 kJ/kg
1 Btu/lb	=	2·326 kJ/kg
Intensity of heat flow rate		
1 $kcal/m^2\ h$	=	1·163 W/m^2 ($J/m^2\ s$)
1 $Btu/ft^2\ h$	=	3·155 W/m^2 ($J/m^2\ s$)
Thermal conductance		
1 $kcal/m^2$ h degC	=	1·163 W/m^2 degC (J/m^2 s degC)
1 Btu/ft^2 h degF	=	5·678 W/m^2 degC (J/m^2 s degC)
Thermal resistance		
1 m^2 h degC/kcal	=	0·859 8 m^2 degC/W
1 ft^2 h degF/Btu	=	0·176 1 m^2 degC/W
Thermal conductivity		
1 kcal m/m^2 h degC	=	1·163 W m/m^2 degC (W/m degC)
1 Btu in/ft^2 h degF	=	0·144 2 W m/m^2 degC (W/m degC)
Thermal resistivity		
1 m^2 h degC/kcal m	=	0·859 8 m degC/W
1 ft^2 h degF/Btu in	=	6·933 m degC/W
Illumination		
1 $1m/ft^2$ (foot-candle)	=	10·76 1x ($1m/m^2$)

Publisher's Note

IN these days it is more and more important for fish to reach the market in good condition. To that end, refrigeration of the catch on the fishing vessel is desirable when its voyage is of more than a short distance from port.

The aim of this book is to establish the principles and practices of sound refrigeration procedure—to make available essential knowledge not only to engineers and operators of the equipment, but to all associated with the handling of the fish so that no diminution of return may be suffered through lack of efficiency.

The author of this work, J. H. Merritt of the Torry Research Station, Aberdeen, was given special permission by the authorities to undertake this work. This he is particularly well qualified to do. The result is a workman-like job which will be invaluable to the technician and the engineer, and also of interest and value to the non-technical reader.

Experience and knowledge gained by practice and research not only in British waters but also in Canada and the United States, have been incorporated. The result is a thorough work which is universally applicable and useful the world over. It will particularly help the fishing industries of developing areas by making available sound principles on which proper refrigeration can be applied to all fishing craft serving their expanding markets.

John H. Merritt was born in Nova Scotia in 1930. After schooling in Halifax he studied mechanical engineering at Saint Mary's University and Nova Scotia Technical College, gaining a Bachelor degree in 1953. This was followed by two years in England, one year at C. A. Parsons & Co. Ltd., Newcastle-on-Tyne, and a year at the School of Thermodynamics, University of Birmingham, which granted an M.Sc.

In 1955 he took up a post with the National Research Council in Halifax, Nova Scotia, working on the drying of seaweeds. In 1958 he was appointed Assistant to the Director with duties in administration and laboratory services, continuing investigations in drying. He lectured part-time in engineering at Saint Mary's University during the session 1959–60.

He joined the Torry Research Station, Aberdeen, Scotland, in 1961 and became Head of the Engineering Section in 1962, working mainly in refrigeration with emphasis on the development of plant and processes for the chilling, freezing and thawing of fish and freezing at sea. The programme of work also has included development of handling and processing machinery and some aspects of fish meal manufacture. He was appointed Assistant Director in 1970. He has contributed important papers to a number of international gatherings on refrigeration and is President of Commission D3, refrigerated sea transport, of the International Institute of Refrigeration. He is a member of the Institution of Mechanical Engineers and the Institute of Refrigeration (UK).

Preface

THE fishing industry has reached a stage where, except in a few areas, fish stocks are being exploited more or less fully. With the help of advanced methods of fish detection and finding, modern fishing gear and the large number of vessels employed, some areas have been fished too heavily. Thus catch quotas and restrictions on catching effort are becoming commonplace. Year by year increased emphasis is placed on conservation of resources. At the same time some attention is being directed toward underutilized species, some of which are difficult to preserve and process. Of course the extension of fishery limits has been a most significant change; it has brought much of the world's fisheries under the jurisdiction of the coastal state. As a result some countries are finding it difficult or impossible to maintain catch rates of the desired species and the larger vessels which fish in distant waters far from the home port have been placed at a particular disadvantage.

In one sense it might be said that catching methods have advanced too far. While the large catch has obvious advantages—even apart from bringing immense satisfaction to the fisherman—it can make proper handling difficult and impose a high refrigeration load, often beyond the capacity of the cooling system. There is, however, much scope for the wider application of good handling and preservation practices which will allow maximum use of the catch and improve the level of quality. Therefore the role of refrigeration will become more important, not only on board the fishing vessel but also in processing, storage and distribution after landing. Present techniques will have to be adapted or new ones devised where necessary.

This book is presented in the hope that it will help in the selection, design and operation of refrigeration systems for fishing vessels, for the benefit of those in the industry and the consumers, especially those suffering from hunger and malnutrition. Much of it also applies to refrigeration ashore. An attempt has been made to explain the methods to the non-technical reader and to concentrate on matters of fundamental importance.

I am grateful to the Director, Torry Research Station, Aberdeen, for permission to publish. Also I am indebted to my colleagues at Torry for their help, especially members of the Engineering Section and J. J. Waterman and to J. Templeton.

John H. Merritt
May 1978

Chapter

Spoilage and Refrigeration *1*

Preservation of the Catch

FISH begin to spoil immediately after death. Efficient methods of preservation on board fishing vessels are necessary in order to land fish of good quality and permit long voyages. Since the rate of spoilage is largely dependent on temperature, increased by increase in temperature, refrigeration of the catch is common practice. Most of the other known methods of preservation including salting, drying, canning, storage under vacuum or gas, irradiation and the application of anti-bacterial substances also have been used or proposed for use on fishing vessels, but refrigeration remains the most popular method and its use is increasing.

Of the methods which do not employ refrigeration, canning, salting, drying and irradiation are the most significant. Canning enables the production of a wide range of products and has been used for sardines, sprats, salmon and other species. It has been practised on board a few large factory ships. Using special packaging materials, a variety of consumer products also can be made in flexible packs instead of cans but this has been confined mainly to shore processing. Salting and drying, which produce considerable changes in flavour and texture, have declined in popularity in many countries largely because improved transport in this century has enabled the rapid distribution of iced fish and other more perishable fish products. Salting has been carried out on board fishing vessels on a large scale but drying has not. Drying plant tends to be bulky and the energy required in drying is greater than in freezing but it could be carried out on board larger vessels with plant of suitable design. Although freeze-drying produces fewer changes than warm air drying, the apparatus is more complex and costly at the present stage of development. Pasteurization of fish by irradiation is of considerable interest but at an early stage of development. Trials have shown that shelf life can be extended but the irradiation plant tends to be heavy and bulky due to the shielding required for the irradiation source.

In this book our attention is focused on preservation methods of practical interest in which refrigeration plays an important part. Broadly, they fall into two categories, chilling and freezing. The main components of the fish to be refrigerated are water, fat and solids, of which a large amount is protein. The solids constitute typically 20 per cent but there are considerable variations. Low values, with correspondingly high water contents, are found in some species and seasonally in fish that have undergone a period of starvation. The amount of fat in the flesh of a lean fish such as cod is low, typically 1 per cent. In some fatty fish the fat content can vary a great deal and reach over 20 per cent in herring, depending on the season and other factors.

It is worth noting that as more mechanization is introduced into the fish industry, the processing characteristics of the fish are being given increased emphasis. Even where delays before processing are fairly short and differences in rate of deterioration are not readily apparent, refrigeration can have an appreciable influence on the degree of success of machine operations. Thus closer attention to handling properties of the fish and to machine output, superimposed on the more traditional criteria for quality assessment, will result in a greater demand for improved refrigeration.

CHILLING

The chilling of fish to a temperature of about 0°C, just above the freezing point of the fish, does not stop spoilage but retards it. Essentially there are three aspects of spoilage in chilled fish; enzymic, bacterial and oxidative changes.

Enzymes are substances present in the flesh and stomach of the fish. They cause chemical changes which during life, but not after death, are counterbalanced with the help of the digestive and blood systems. Usually there is food in the gut of the fish when it is caught and powerful digestive enzymes are present. The gut wall and the neighbouring flesh are readily penetrated and softened by the enzymes after death, even in chilled fish, and thus easily invaded by spoilage bacteria. It is the custom, therefore, in the trawl, seine and line fisheries, for demersal fish such as cod and flatfish to be gutted and washed after catching. Fish generally are not gutted in pelagic fisheries where small fish often are caught in greater numbers, hence spoilage tends to be quicker. Enzymic activity is probably largely responsible for the flavour changes that take place in gutted fish during the early period of storage when there is no marked bacterial spoilage. The process of *rigor mortis* or death stiffening, in which enzymes play a part and the intensity and duration of which depends on temperature and on the species and condition of the fish, largely passes off before bacteria begin to penetrate and develop in the flesh.

Bacterial action, aided by the changes caused by enzymes, is by far the main cause of spoilage in chilled fish. Bacteria in large numbers are confined to the surface slime, gills and intestine of the live fish. These organisms, usually together with bacteria introduced from other sources, increase in numbers after death and eventually penetrate through the skin and invade the flesh.

Rancid odours and flavours are produced when oxygen in the atmosphere combines with fat in the fish. Oxidation does not have a marked effect in lean fish but can play a significant part in the spoilage of fatty fish such as herring and mackerel.

These changes are influenced by factors other than temperature and, altogether, many factors have a bearing on the final quality. For example, the condition of the fish as caught can vary a great deal with the season, fishing ground, etc., and quality will be affected by the method of fishing and the techniques employed for bleeding, gutting, chilling and storage. Different kinds of fish will spoil at different rates and in different ways. Small fish tend to spoil more rapidly than large fish, largely because surface bacteria can exert more influence. Shellfish can become severely discoloured. The characteristic pink colour in some shrimp is caused by the initial spoilage processes and advanced spoilage can be accompanied by black discolorations. Nevertheless, temperature is the most important single factor affecting the spoilage rate and the main requirement of any system of chilling will be to maintain the lowest possible temperature short of freezing.

According to normally accepted standards, cod stored at a temperature of 0 °C in melting freshwater ice has a keeping time of about 14 days, small haddock slightly less, hake a good deal less and flatfish slightly more. Ungutted herring in ice has a keeping time of roughly six days, except during periods of heavy feeding when the time can be much reduced. Sardines have a shorter storage life than herring. On the other hand large tuna and halibut have a storage life of more than 20 days. Large shrimp in ice have a storage life of several days. As shown in Fig 1.1, the cod is considered to be very fresh for only the first seven days of storage in ice. Much of the iced cod discharged from the larger trawlers, which often return to port incompletely filled because of the spoilage limit of the early part of the catch, has been held in ice for more than seven days.

These interdependent problems of quality and length of voyage are widespread. They must be considered along with the question of

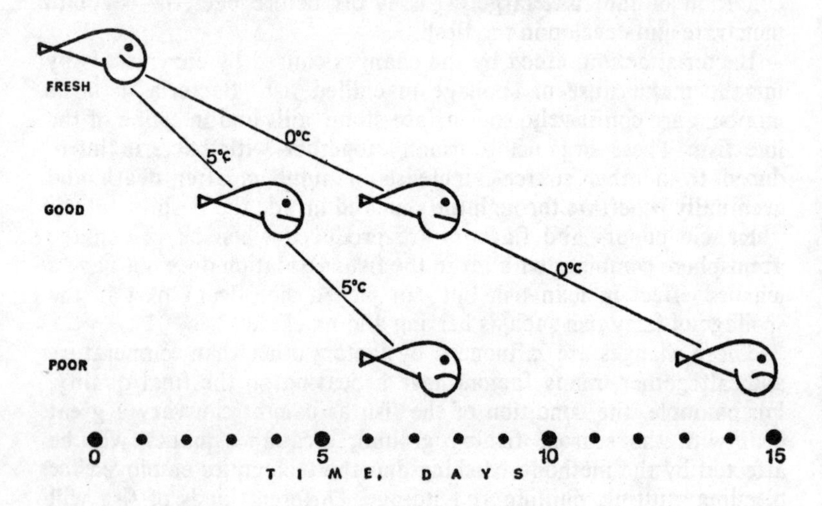

FIG 1.1 Spoilage of cod

efficient use of labour and investment. There has been, therefore, a search for suitable methods of chilling which would reduce the rate of spoilage below the rate in melting ice, the most common method. Even in fisheries where the length of voyage is somewhat short of the limit imposed by chilling in melting ice, a further reduction in the rate of spoilage is desirable, bearing in mind the difficulties in handling and distribution on shore. At present, however, the alternatives to melting ice are limited. The most important is refrigerated sea water which in many cases gives advantages, mainly in the handling of the catch.

FREEZING AND COLD STORAGE

The spoilage processes which limit the length of time of chilled storage are virtually stopped by quick freezing and cold storage at −30 °C or below. At −30 °C in a properly designed and operated cold store, gutted cod can be kept for eight months and ungutted herring for six months without significant change. The stored fish will be very nearly as good as fresh. As in chilling, the temperature of storage is most important. The period without significant change will be reduced to four months for cod and three months for herring at a cold store temperature of −20 °C. At −10 °C the period will be reduced to one month in both cases. Fish in poor condition before freezing will have a reduced storage life.

Whereas pure water freezes at 0 °C, the water in the fish does not begin to freeze until it reaches −1 °C or below because of the other substances present. As the temperature is reduced, more and more water is frozen. It is worth noting that, although bacterial action is progressively reduced by temperatures below 0 °C and practically ceases at −5 °C, the salts and other chemicals including enzymes increase in concentration in the unfrozen water as more and more of the water in the flesh becomes frozen. By this means their activity can be relatively high just below the point where freezing begins, in the region of −1 to −5 °C, associated with changes in the protein.

There is no doubt that mechanical stresses are produced during freezing due to the dimensional changes which occur when the water in the flesh is frozen. Little study has been made of this phenomenon but some aspects are discussed in Chapter 8.

The main changes that occur during cold storage are protein denaturation, oxidation and dehydration. Freezing itself causes some protein changes (denaturation). A notable exception to the rule that a storage temperature of −30 °C is low enough is the tuna, which is stored at temperatures below −35 °C in some markets in order to prevent rapid changes in the colour of the flesh, due to oxidative changes during storage.

The protein changes in cold storage, if excessive, cause the flesh of the thawed fish to be spongy and dull in appearance. Undesirable flavours are present. Juice or drip tends to run out and can be squeezed out easily. Badly denatured fish does not make a good smoke cure, largely because the surface lacks the gloss typical of good quality smoked fish. Drip in white fish fillets can be reduced by dipping them in brine or other solutions before freezing but dipping at the time of thawing is also effective and usually preferred. The main concern on board ship will be to quick-freeze and store the fish at −30 °C or below so that the changes will be at a minimum.

Rancid flavours can develop in fatty fish because the fat readily combines with oxygen. The rate of oxidation is largely dependent on temperature but it is influenced by other factors. It is assisted by the enzymes present in the fish and can be accelerated by the presence of salt and other chemicals. Glazing of the fish by dipping briefly in water or by brushing or spraying on water provides some protection against oxidation. So does vacuum packing in bags of suitable material such as some plastics, impervious to moisture and oxygen, which has been used for small consumer packs. The storage life of whole fatty fish such as herring and mackerel, frozen at sea in blocks, can be doubled by freezing them in water in protective bags. Another approach is to incorporate antioxidants, chemicals which retard oxidative reactions in the fish, in the glazing or in a solution applied to the fish before freezing. So far there has been little success in this direction. Again, the emphasis on board will be on quick freezing followed by storage at −30 °C or below.

Some dehydration of the fish can take place in cold storage. There also can be some loss of moisture on quick freezing but normally it has a negligible effect on quality. Excessive drying encountered in cold storage alters the appearance of the fish, making it dry and white, an effect known as 'freezer burn'. It is often accompanied by high rates of protein denaturation and oxidation. Many factors influence the rate of drying but underlying all of them is the fact that heat is required for evaporation. All heat gains to the cold store are potential causes of dehydration and must be given attention in cold store design and operation. Sometimes impervious wrappers are used to prevent dehydration but some transfer of moisture can take place within the pack itself. Water glazing also is used to protect the fish and is common practice with unwrapped fish. The glaze evaporates during storage and may have to be renewed after a time. Often the fish are not glazed at sea but are glazed after discharge for cold storage on land, especially when the period of cold storage on board is short and the storage conditions are good. Fresh water should be used for glazing.

It is worth emphasizing that freezing and cold storage inevitably will cause some deterioration; the product will not be improved under any circumstances. Carried out properly, however, the method provides almost perfect preservation.

THERMAL PROPERTIES

Whatever the refrigeration system, its function is to reduce the fish temperature as necessary, usually quickly, and then maintain the required temperature against the ingress of heat. In order to reduce the temperature, heat must be extracted.

Fish temperature and heat content

The relationship between fish temperature and heat content is given in Fig 1.2. The temperature of −30 °C has been chosen arbitrarily as the zero point where freezing is complete. Strictly speaking, however, not all the water in the flesh is frozen at −30 °C or even at −80 °C but about 80 per cent is frozen at −5 °C and nearly all at −20 °C. The approximate specific heat of the fish is 4·0 kJ/kg degC above 0 °C and 1·6 kJ/kg degC below −20 °C.

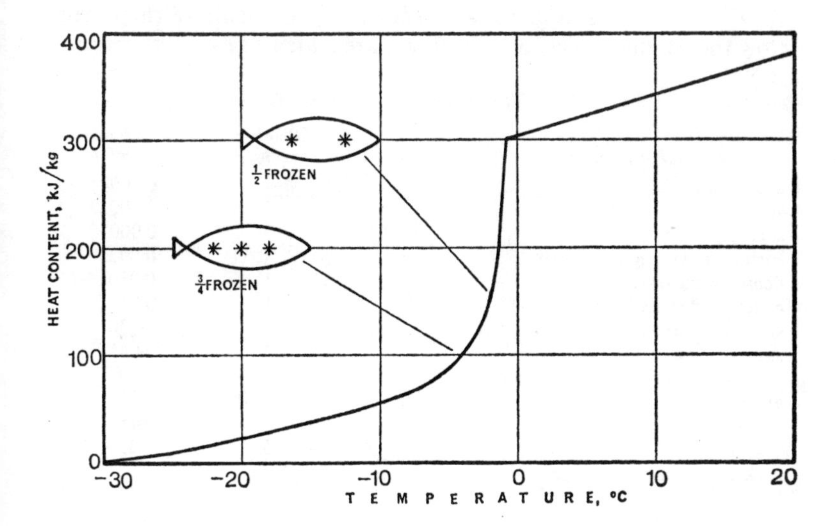

FIG 1.2 Heat content of cod

This data is for white fish muscle with a moisture content of 80 per cent and is reasonably accurate for whole white fish. The heat content of fatty fish is slightly lower and freezing occurs at a lower temperature level, depending on the fat content. Thus fish with a high fat content of about 20 per cent may begin to freeze at $-2\,^{\circ}$C instead of $-1\,^{\circ}$C as shown and the heat content may be 22 per cent lower. For most purposes, however, the estimation of cooling loads based on the data in Fig 1.2 is sufficiently accurate.

Resistance to heat flow

The heat flow rate through any body will depend on the properties of the body such as type of material, size, shape and surface reflectivity. It will depend also on the temperature difference across it and the area involved. Therefore it is convenient to consider the heat flow rate through a unit area for a unit temperature difference. This rate is called the *thermal conductance* which defines the ability of a body to conduct heat and can be expressed in W/m^2 degC. The reciprocal of the thermal conductance, m^2 degC/W, is called the *thermal resistance*. This is a useful term when estimations of heat flow rate and comparisons of various bodies or structures have to be made. According to the basic laws of heat transfer, resistances can be added to calculate the heat flow in composite systems with series flow of heat. With parallel flow the conductances, or heat flows, are added.

Table 1.1 Typical Values of Thermal Resistance

Material or item	*Thickness, cm*	*Resistance, m^2 degC/W*
aluminium	1	0·00005
steel	1	0·00023
surface, water at 1 m/s	*	0·00050
surface, mildly agitated water	*	0·0025
dense plastic or rubber	1	0·010
concrete, density 2 t/m^3	1	0·013
surface, air at 7·0 m/s	*	0·030
surface, air at 3·5 m/s	*	0·045
surface, natural convection of air	*	0·10
air space	2 to 10	0·17
wet wood	5	0·18
dry wood	5	0·33
cork insulation	5	1·4
frozen fish	5	0·03
unfrozen fish	5	0·09

* surface resistance; thermal resistance at a solid boundary, reciprocal of the surface heat transfer coefficient.

Where there is heat transfer between a solid body and a fluid such as air or water in contact with it, there is a resistance at the boundary due to a layer of tranquil fluid. The depth and thermal resistance of the layer will depend on its characteristics including viscosity and the amount of agitation or the velocity in the bulk of the fluid. The term *surface heat transfer coefficient* is often used to define the thermal conductance in this case.

Some values of resistance are given in Table 1.1. They are only approximate, to serve as a guide and illustrate orders or magnitude. They should not be used where accuracy is essential, especially where total resistance is low. Surface resistance is dependent on a number of factors because conduction, convection and radiation all are involved. It can be very low where there is condensation of vapour on a cold surface. The values given for an air space and surfaces bounded by air are roughly correct for moderately smooth non-reflective surfaces. Highly reflective surfaces bounded by air will have resistances of more than twice the values listed. A moisture content of 50 per cent of the dry weight has been assumed for wet wood but the values for wood also can vary according to the species.

Chapter

Icing 2

STOWAGE IN ICE

IT is not surprising that freshwater ice has played and continues to play a major role in the chilling of fish on board because it has several advantages over other forms of refrigeration. It involves little or no complication in the design and operation of the fishroom or storage space. It provides clean, moist, aerated storage and, since the temperature at which the ice melts is fixed, there is some degree of control of fish temperature. Given good contact between fish and melting ice, cooling is quick. The ice also can cope with heat from external sources and whatever heat is generated by the spoilage processes.

Temperature

Provided the fish is completely surrounded by melting freshwater ice, it will be chilled to about 0°C, the melting point. In fact, fish temperatures slightly below 0°C have been recorded in many instances in trawlers because the melting point of the ice is reduced slightly by the salts, blood and slime present in the catch. Using ice alone, however, the fish temperature cannot be reduced to the point at which freezing begins, between −1 and −2°C. According to the best technique, therefore, each fish should be in intimate contact with melting ice so that fish temperature is reduced as quickly as possible and maintained as low as possible.

An indication of the effectiveness of the method of chilling can be obtained by measurement of the temperature of the fish during storage, usually by insertion of a thermometer probe into the fish. The technique is discussed in Chapter 8. A typical record of temperature with good icing, using a thermocouple at the centre of a gutted cod 3 kg in weight is shown in Fig 2.1. The ambient temperature was 2·5°C and the fish was one of several iced in an aluminium box with a capacity of about 100 dm^3. The initial cooling to 0°C is rapid, approximately 6 h in this case, nearly as rapid as in cold water.

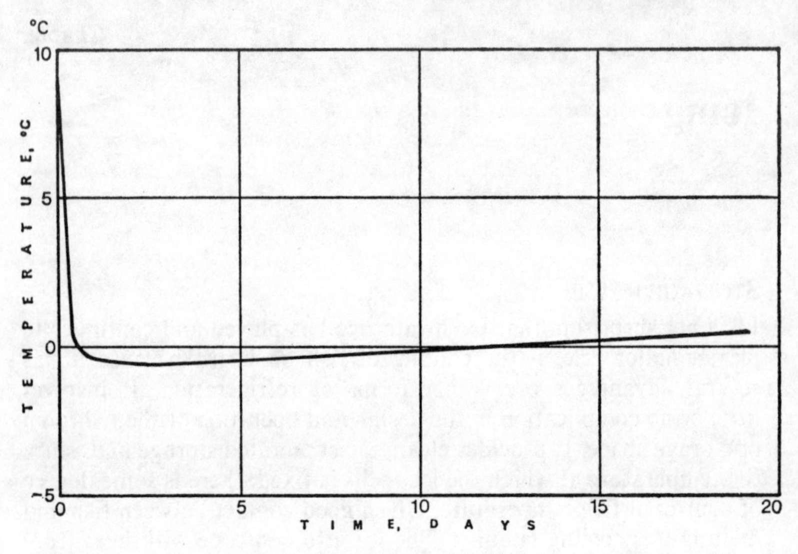

FIG 2.1 Temperature of iced fish

Meltwater

The cooling of the fish is largely achieved by the cold meltwater flowing over it. The melting rate is higher when the fish are warmer, so cooling is applied automatically and quickly but without the danger of freezing. In addition to good chilling conditions, the meltwater assists in the removal of blood, bacteria, etc. This accounts for the slight increase in fish temperature during storage, shown in Fig 2.1. In many markets blood removal is desirable from the point of view of the appearance of the flesh. It is essential to have good drainage in the fishroom so that the fish do not lie in contaminated meltwater.

Depth of stowage

The depth of fish and ice should be limited in order to avoid crushing and weight loss in fish due to the weight of the bulk. Also, with limited depth and adequate drainage, exposure of the lower fish to excessive amounts of contaminated meltwater from above is avoided. With most species the depth should not exceed 40 cm but for some fish, for example herring, the limit should be less if damage due to pressure is to be avoided.

Cleanliness

Cleanliness is essential. Fishroom, fishroom fittings and containers should be designed so that they can be cleaned effectively and easily and they should be maintained in clean condition.

Species which can discolour other fish and those which produce a considerable amount of ammonia during spoilage, such as skate and dogfish, should be segregated from the rest of the catch.

Ideally then, each fish should be surrounded by ice. This ensures that the catch is not only cooled as rapidly as possible and maintained at 0°C but that the mass of fish and ice is ventilated. If the fish are placed against one another or against other smooth surfaces such as pound boards or sides of the fishroom there is a risk of serious spoilage characterized by foul odours, sometimes called bilgy odours, due to the exclusion of air.

Labour

Stowage on board with proper icing can involve a great deal of labour, especially in the handling of fish and ice. In many cases the only mechanical aids to the tasks of gutting and icing are the gutting knife and the shovel. Gutting machines have been introduced in recent years with some measure of success but there is a need for machines of greater reliability, suitable for large and small vessels and a variety of fish. Easier methods of dispensing ice, often with the aid of a blower which transports the ice in a stream of air, are being applied in some fisheries. Considerable effort is being directed toward these problems and there is hope for further progress.

Easier handling, higher stowage rate and possibly other advantages along with an acceptable spoilage rate are sought in alternative methods, notably refrigerated sea water, particularly in some pelagic fisheries where the fish are caught in large numbers and stored ungutted for relatively short periods.

THE ICE

The water used for ice manufacture must be fit to drink. Even though ice is made from clean, potable water, however, appreciable numbers of bacteria can build up in the ice depending on the temperature and length of storage. Thus old ice in a wet fishroom will be heavily contaminated with spoilage bacteria. Spoilage of the fish will be more rapid in old or dirty ice, so every trip should be started with fresh ice.

There are three common forms of ice according to the method of manufacture: crushed ice, flake ice, tube ice. There is little to choose between these various forms on the grounds of quality. The important thing is to use ice!

Crushed ice

Crushed ice is made from blocks and slabs. This method has been employed on a large scale in large ports. Block ice can be crushed to any desired degree of fineness but commonly it consists of irregular lumps 6 mm to 50 mm in thickness. A disadvantage is that the large pieces sometimes make indentations in the flesh of the fish. The normal bulk density is 640 kg/m^3.

Flake ice

Flake ice is made by freezing water in thin layers on a smooth refrigerated surface. The ice is removed by mechanical action, for example by a scraper on a cylindrical surface, or by a hot defrost. Ribbon ice, a form of flake ice, is made on a flexible refrigerated pipe and peeled off.

Flake ice has the advantage of small particle size. Typically the flakes are 3 mm thick with a slightly curved area of 6 cm^2. Flake ice has a lower bulk density than crushed ice, about 480 kg/m^3, so that the volume of ice is greater for a given amount of cooling. This is not a serious disadvantage in most cases.

It might be expected that flake ice would give relatively quick cooling because of better contact with the fish but it is doubtful whether there are any significant differences in the cooling rates with various forms of ice. Because of its lightness and shape of the particles, flake ice sometimes forms bridges as it melts, leaving a gap at the top of the fish to be cooled.

Flake ice machines have been used on board ship for the manufacture of ice from sea water, particularly on factory vessels for the chilling of the catch before further handling, or freezing on board.

Tube ice

Tube or cylindrical ice is formed inside a refrigerated tube, removed with the aid of a hot defrost and cut into lengths. Typical dimensions of the pieces are 40 mm in diameter with a hole 10 mm in diameter and 40 mm in length. The bulk density is about 550 kg/m^3, midway between that of flake ice and crushed ice.

Cooling capacity

The store of ice in the fishroom should be drained so that what is used is ice, not a slushy mixture of ice and useless water. Storage of the ice at temperatures below 0 °C will keep it dry and loose, prevent loss by melting and prevent fusion of the ice into a mass at the surface which makes handling difficult. Over longer periods of storage it will also significantly retard the development of bacteria in the ice. The cooling capacity of the ice will not be increased greatly, however, by a reduction in temperature. The heat absorbed by each kg of ice on melting at 0 °C is 333 kJ. The heat absorbed on melting from an initial temperature of −5 °C is 343 kJ/kg, an increase of only 3 per cent.

When using or purchasing various kinds of ice it is important to remember that the cooling capacity depends on the weight of ice although the measure may be volumetric. For example, four shovelfuls of flake ice are equivalent to three shovelfuls of crushed ice.

Fishroom Conditions

It has been shown in laboratory tests with carefully iced fish that a high ambient temperature, say 15 °C, which is accompanied by a high rate of melting, can give marginally better results than a low ambient temperature, say 2 °C, with a low rate of melting. In practice, however, the deliberate use of higher temperatures usually is not feasible for several reasons. With inadequate icing or insufficient amounts of ice, the heat introduced would greatly increase spoilage rate whereas the consequences of imperfect icing at low ambient temperature would not be so severe because fish temperature would remain low. Also, usually it is not possible to ice the catch more than once. High melting rate requires a greater quantity of ice, hence greater cost, and the stowage rate of the fish is reduced.

Heat gains

In order to have good storage conditions throughout the period of the voyage it is essential that the fish be protected from any heat which enters the fishroom. Insufficient protection will result in undue spoilage and, under some conditions, desiccation of the fish. Ice should be used to absorb heat gains. Accordingly, depending on the rate of melting, 15 cm or more of ice should be placed at the bottom and sides of the fishroom and at the top of the stowed fish on the longer voyages. Experience is the best guide in determining the amount of ice required. When the fish is discharged at the end of the voyage some ice should remain at the surfaces of the fishroom. The melting rate can vary enormously as it is dependent on outside temperature and the resistance of the fishroom to the ingress of heat. Heat gains through the sides of a poorly insulated fishroom can be considerable, corresponding to a melting rate of ice of more than 10 kg/day per m^2 of surface, or 2 cm/day, with high outside temperature. In some modern vessels with good fishroom insulation the heat gains through the insulation are small even at high outside temperature. Nevertheless the fish must be protected by an adequate layer of ice at the inside surfaces. Very often it is necessary to place emphasis on bulkheads, especially those separating the fishroom from the engine room which usually is at high temperature. Of course heat will enter the fishroom through uncovered surfaces. These gains will tend to heat the air in the room and eventually be absorbed by the ice.

The ice must cope with a number of heat gains apart from those carried through the structure. Lights will contribute some heat, so should not be left on unnecessarily. A man working in the fishroom will contribute roughly 0·30 kJ/s. Also, the action of the spoilage bacteria produces a significant amount of heat. The most important, however, is heat added by the introduction of outside air through openings in the fishroom. The rate of air change in the fishroom should be kept low by restriction of the openings. More than one opening is undesirable because it leads to a high rate of air change. The hatch should not be left open any more than necessary during the voyage. This is particularly important in small fishrooms where the influence of the hatch opening is relatively great.

Insulation is installed in fishrooms in order to reduce heat gained by the fishroom, in other words to reduce the melting rate and prevent spoilage due to high temperature. The same factors have largely contributed to the trend toward the use of mechanical cooling in the fishroom as a supplement to ice, particularly in warm climates. Insulation and mechanical refrigeration are discussed in Chapters 4 and 5.

TREATMENT OF THE FISH BEFORE STOWAGE

Bearing in mind the dependence of spoilage rate on temperature, it is imperative to chill the catch as soon as possible after it is landed on deck. In some cases, for example with hake in a warm climate, a delay of 2 h may be excessive. Even in arctic waters the fishing deck can be warm, particularly in direct sunshine. Rough treatment of the fish should be avoided because bruises and unnecessary cuts in the fish only aggravate the spoilage problem by causing discolorations in the flesh and allowing easier entry of spoilage bacteria. Clean conditions should prevail so that the numbers of spoilage bacteria and undesirable material introduced into the fishroom are held to a reasonably low level. The deck, baskets, boots and other items in contact with the catch should be thoroughly washed by hosing before fish are landed on deck. Cleaning normally should be carried out immediately after each haul of fish has been dealt with. One haul should not be dumped on top of another because it will lead to mixing and can mean an unduly long delay before stowage for some fish. In a heap of fish the spoilage rate can be accelerated because of the exclusion of air and spontaneous increase in temperature due to bacterial action.

Shellfish can be iced in the same way as other fish and processed ashore. In some fisheries they are cooked in boiling sea water on board.

Gutting

Gutting, removal of the guts, is not always carried out. Rough and spiny fish are awkward to gut by hand and so sometimes are left ungutted. Redfish fall into this category but, fortunately, because they are caught in deep water the contents of the stomach are discharged by virtue of the decrease in pressure on bringing them to the surface; thus the need for gutting is not so great. When fish are caught in large numbers, particularly small fish, it may be impossible to gut.

Generally speaking, spoilage and discoloration during storage will be greatly decreased by gutting. Early gutting and the avoidance of increased temperature followed by storage in melting ice will effectively bleed the fish, leaving the flesh free of blood discolorations. In order to prevent needless contamination, gutting should remove all the gut and liver. Guts and unwanted fish should not be mixed with fish destined for stowage. Excessive and ragged cuts should be avoided but the belly cavity should be opened sufficiently to make washing after gutting effective. The cutting of both napes of roundfish such as cod helps to achieve this.

It has been standard practice to gut the fish with the aid of only a knife. Much of the effort expended by the fisherman, who is often exposed to the elements, is in stooping to pick up the fish and then throwing it into the washer or washing it manually after gutting. The gutting rate will vary according to the type of fish. The usual rate for cod is about three fish per man-minute or roughly 360 kg/man h with fish weighing 2 kg. More civilized aids such as shelter decks and gutting benches to which the fish are fed at waist level have been introduced. The gutting bench increased the gutting rate by about 25 per cent and made the work of gutting easier but the quality of gutting may not be as high. Gutting machines have begun to come into use and there is some promise that simple and reliable machines will take over much of the work. The gutting bench and the gutting machine lend themselves to conveyor systems which also can reduce handling and damage to the fish through rough handling.

Heading

Heading, removal of the head, is sometimes practised in order to increase the stowage rate of edible material. Also some types of gutting machines include heading as an initial step before gutting. It should be carried out by cutting, at least through to the bone, not by tearing off the head. Heading may result in a loss of yield on later processing but if it is done accurately and cleanly any loss will be slight. Where the fish is to be filleted for example, there may be losses of material on heading and on filleting, because the exposed end of the fillet may be discoloured and have to be trimmed off. Large shrimp often are headed before stowage.

Bleeding

Thorough bleeding of the fish is often essential. This is done in some fisheries by cutting the throat of the fish soon after catching, before gutting. Heading or gutting in the recommended manner with adequate chilling is also effective if carried out early enough. Chill conditions, below 5°C, should prevail before and after the cutting operation for best results if blood discolorations are to be avoided, in order to prevent clotting of the blood. The time required for adequate bleeding varies considerably depending on the condition of the fish and on temperature. In most cases 1 h in ice or in chilled water is sufficient. With proper handling and stowage, blood discolorations will be at a minimum because stowage in ice provides good conditions for bleeding.

Washing

The fish usually are washed before stowage although it is not always necessary. When the catch is visibly dirty, it is advisable to wash the fish by hosing them with water. It is normal practice to wash the fish in sea water after gutting in order to substantially eliminate loose dirt, blood, intestinal material, etc., from the outer surfaces and belly cavity of the gutted fish. There are various methods of washing. When the fish are washed by hand in batches with the aid of a hose it is well to avoid a long delay between batches in order to stow them as soon as possible after gutting and promote more effective washing in small batches. Sometimes the fish are placed in open mesh baskets and lowered into a tank of agitated water.

In another method the fish are tossed into a bank of swirling water, supplied through jets, immediately after gutting. Assisted by the motion of the ship they then pass over a weir at one end of the tank and down a chute into the fishroom. This type of continuous washer was widely adopted in the 1950's for the larger trawlers. It virtually eliminated delay between gutting and stowage and smoothed out the flow of fish to the hold. Other types of continuous washer have followed. One consists of a rotating metal cylinder roughly 7 m in length, installed at a slight slope so that the fish pass through. It is fitted with water sprays and drainholes inside.

In all the methods of washing, the time spent by the fish in the washer is short, usually less than 3 min.

BULKING

In bulking, the fish and ice are mixed to achieve intimate contact which will secure the maximum storage life in ice. It usually is carried out in a hold divided into pounds fitted with removable shelves of wood or metal. Bulking to a depth greater than 40 cm is not recommended for cod because it can lead to damage and severe weight loss due to pressure. With some other species the limits are lower. Weight losses in excess of 10 per cent over a period of 14 days have been recorded in cod and haddock stowed at a depth of 1 m. Care must be exercised, therefore, to ensure that the shelves are not overfilled and that each is resting on its supports, not on the bulk of fish and ice immediately below. As in any method of icing, the fish should be protected by ice against the ingress of heat. Shrimp will have a relatively high weight loss, perhaps more than10 per cent in some instances, even with good icing.

With the method of bulking, there are difficulties in the unloading of the catch. A large amount of labour is required and there appears to be limited scope for mechanization. The most common procedure is to separate the ice and fish on board, load the fish into baskets or boxes and then transfer them to shore with the aid of a winch. The ice is discharged overboard. Alternatively, the ice and fish can be separated on shore. Often the fish are damaged by hooks and shovels and lie on the market for several hours without any ice. Conveyor systems, some of which automatically separate fish and ice on landing, offer some improvement and are coming into use. According to another proposal, the fishroom would be made up of one or more removable compartments which could be lifted out and emptied by tipping. Pumps also have been used but in some applications there has been physical damage to the catch. There are systems where the fish are mixed with a substantial amount of water and pumped out by a centrifugal pump. Another pumping system employs a vacuum with only a little water to ease the movement of fish. Pumping systems are widely used, without appreciable damage to the fish, for pelagic fish to canneries and have been used for fish such as menhaden and anchovy, uniced and destined for fish meal manufacture. There has been limited success with the larger fish.

For the purpose of quality control, it is essential to have a good stowage plan in order to distinguish between fish held for various periods of storage before landing and fish of various species and sizes. The method of bulking, however, makes it difficult to avoid mixing of the catch at the time of discharge.

Bulk stowage without ice

Bulk stowage without ice is sometimes practised where icing is thought to be unnecessary or too difficult, for example with pelagic fish caught in large numbers. The storage life is dependent on temperature, species and other factors but in any case is relatively short.

SHELFING

Shelfing means the stowing of fish in single layers, gut cavity down, on a bed of ice. Sometimes a little ice is spread on top. It is designed to ensure bleeding through the cut surfaces of the fish and retard spoilage by chilling, bearing in mind that the gut cavity can be a source of trouble. The skin of cod stored in this way has a glossier, more attractive appearance than the skin of bulked fish. There is an absence of indentations caused by ice which may be present in bulked fish. Shelfing is a more laborious method of stowage and consumes about double the space required for bulking. It also produces inferior quality because icing is incomplete, giving higher fish temperature and more influence from fishroom conditions. With the usual fishroom temperature in arctic fishing, between 1 and 2°C, significant differences in quality between bulked and shelfed fish can be expected after 3 to 7 days' storage.

In some fisheries it has been the practice to shelf cod and haddock toward the end of the voyage but in recent years as catching rates have declined the proportion of shelfed fish has increased. The demand for shelfed fish has been due to its better appearance and the fact that it is associated with the latter part of the catch. In fact the term 'shelfing' is being used more loosely than before, to describe bulk stowage in ice in shallow layers.

Sometimes live crab and lobster are held by placing them on a layer of ice with adequate ventilation and wetting.

Boxing at Sea

Although bulk stowage in ice can be just as good from the point of view of quality landed, boxing in ice at sea has several advantages over bulking. Discharge is made easier and lends itself to mechanization. Boxing can eliminate handling of the fish on shore and consequently improve quality at the point of consumption. Whereas there is often a lot of mixing of the catch on discharge with the other methods, particularly with bulked fish, boxing simplifies the problem of segregation of first caught from last caught, small from large and one species from another. In many markets, however, the traditional methods of buying on inspection and by auction demand that the entire catch be displayed without ice and that the weight of fish be known fairly accurately. Thus the fish suffer serious increased spoilage due to increase in temperature and additional handling and delay. Suitable devices for weighing at sea could help to eliminate this problem but are not available. This difficulty is avoided by other methods of buying such as purchase by contract, still retaining a free market system.

A disadvantage in some cases is that the space occupied in the fishroom is greater than for bulking, usually not more than one-and-a-half times. The weight of the box when filled is normally between 30 and 65 kg depending on the desired capacity. Apart from having the proper shape and dimensions for the fishery concerned, the boxes must be robust, easily handled, filled and emptied, easily cleaned and stack securely one on top of the other. There should be adequate provision for drainage of meltwater from the box. Blockage of drainholes at the bottom of the box should be avoided by good design, cleanliness and by placing enough ice at the bottom to prevent blockage by the fish. Ideally each box should drain outside the box below, not into it. Usually some sort of nesting feature with the empty boxes is required in order to make room for ice and leave working space in the fishroom. Cleaning can be carried out at some central point ashore. Although wooden boxes are often used for boxing at sea, boxes of metal and of plastic are popular largely because of their durability and ease of cleaning.

The boxes should be designed with no sharp corners or inaccessible areas which would make cleaning difficult.

Direct contact between the surfaces of the box and large areas of fish should be avoided. This imposes no great difficulty. Much as in bulked stowage, ice should be placed at the bottom of the box. Each layer of fish should be covered with ice before the next layer is placed in the box and then ice should be placed on top. The box should not be overfilled as this will make stacking difficult and put pressure on the fish, possibly leading to a loss of weight during storage. Cod stored in ice in boxes normally gains slightly in weight, 1 to 2 per cent, over a period of 14 days.

In addition to the ice used in the boxes, ice should be placed outside the boxes to absorb heat gains as described above. The boxes should not rest directly on the ice at the bottom of the fishroom but on battens or similar supports, clear of the ice. The fishroom drainage system must function as with the other methods of stowage.

The filled boxes usually are stowed in horizontal layers or in steps vertically so that there is no danger of instability as the stowage increases in height in the fishroom.

The better boxes are expensive even when mass produced. In order to gain the maximum benefits, a large number of standardized boxes must be used on a wide scale and the marketing system should be such that the boxes are not lost or taken out of use.

A common fault is to use insufficient ice because the box is not large enough. Sometimes, when the amount of fish to be stored in the box is more or less fixed, there is not room for enough ice. Particular care should be taken when there are warm fishroom conditions. In this respect the insulation value of the box material may be of some importance. The rate of depletion of ice will be high in boxes of wet wood and of dense plastic and will be highest in metal boxes. By using boxes made of insulating material, icing can be made more effective and more independent of fishroom conditions. An insulated box also has advantages where fish is held or transported in the box on shore after landing. Handling of the fish, which in itself causes some loss of quality, and re-icing can be eliminated. Boxes of plastic foam between two thin layers of dense plastic and possibly other types may prove suitable for boxing at sea if they can be produced cheaply enough. Small boxes of expanded polystyrene plastic have been introduced for the transport of iced fish products on shore but they are not suitable for boxing at sea. Of course the alternatives are to use more ice, outside the boxes as well as inside, and install a mechanical refrigeration system in the fishroom. Mechanical systems with icing are discussed in Chapter 5.

In one method of icing in boxes, strongly recommended, a form consisting essentially of four vertical walls of rigid sheet is installed in the box before filling with fish and ice. The form is designed to give an outer space of 50 mm or so for ice only at the sides of the box and an inner rectangular space for fish iced in the usual way. It is removed after the box is filled.

Transfer at sea

Transfer of fish from one vessel to another has been proposed and practised from time to time as a way of improving the productivity and efficiency of a fishing fleet. It may be a question of transfer from catcher vessels to a factory ship or transfer between two fishing vessels in order to spend more time on the fishing grounds by making better use of available storage space. At the same time, the period of storage can be reduced, resulting in landings of better quality. Some methods have employed small boats on the open sea, carrying the catch from one larger vessel to another, often involving a considerable hazard for the fisherman. The fish can be left in a net in the sea by the catcher vessel with a suitable marker or other means of locating it but, unfortunately, the fish can deteriorate rapidly under these conditions. Sometimes severe changes occur in 1 h. Various direct methods such as transfer by flexible pipeline and pump, floating containers, tackle runners, etc., have been employed.

Boxing at sea is a suitable method of stowage where fish are to be transferred from one ship to another at sea. Direct methods for the transfer of iced fish in boxes have been developed but not been practised on a very wide commercial scale. Using inflated fenders, ships of various sizes can be berthed together and boxes transferred by derrick in rough weather.

AMOUNT OF ICE

The amount of heat removed from 1 kg of fish on cooling from 20°C is about 80 kJ. Ice absorbs 333 kJ/kg on melting, so 0·25 kg of ice will be required for only the cooling of the fish. With proper mixing of fish and ice this melting will occur in a few hours, as demonstrated by Fig 2.1, even with fish weighing substantially more than 3 kg. Thus it may be necessary to employ more than 0·50 kg of ice per kg of fish, not including ice required to cope directly with heat gains from outside.

The correct amount of ice to be used also will depend to some extent on the duration of the voyage and the rate of melting. For storage periods of up to 14 days under arctic conditions white fish should be stowed with a fish to ice ratio of not more than three to one by weight, not including ice used to cope with heat gains from outside. Thus the overall ratio normally is about two to one. Under tropical conditions the ratio is reduced to as low as one to one for the longer periods of storage, half the arctic figure. Greater amounts of ice are required in warmer waters in order to cope with the large initial cooling load imposed by the warm fish as well as increased cooling load imposed by the warm ambient conditions.

Stowage Rate

The stowage rate of the fish in ice will depend somewhat on the density of the fish itself and the amount and density of the ice. The bulk density of the fish varies from 800 to 1,000 kg/m^3 depending on the amount of fat in the flesh, firmness, whether it is gutted, etc. The bulk density of flake ice is three-quarters that of crushed ice and tube ice lies about midway between the two. Some values of stowage rate, based on a bulk density of 900 kg/m^3 for the fish, are given in Fig 2.2. Allowance has been made for the small amount of space occupied by the pound structure in bulking.

High stowage rate can be an advantage where fishroom size is limited. In many fisheries the length of voyage is not restricted by the size of the fishroom, however, but by the spoilage factor, market conditions and other considerations. Even in cases where the size of the fishroom does limit the voyage, increased stowage rate at the expense of adequate icing cannot be a sound proposition in a market that places a value on quality.

The stowage rate of edible material can be increased by removing and discarding unwanted parts of the fish, although often they are of some value for the production of fish meal. With cod, for example, there is a gain of more than 30 per cent by removal of the heads.

Standard of Icing

In some fisheries, in order to permit as long a voyage as possible, icing on board is carried out efficiently and closely approaches the ideal. Often practice falls far short of the ideal, however, particularly on relatively short voyages. Thus, while the fish may be of acceptable quality when landed at the end of the voyage, shelf life and quality at the point of consumption will be lost. When catching rate is high it may be impossible to gut the fish and the standard of

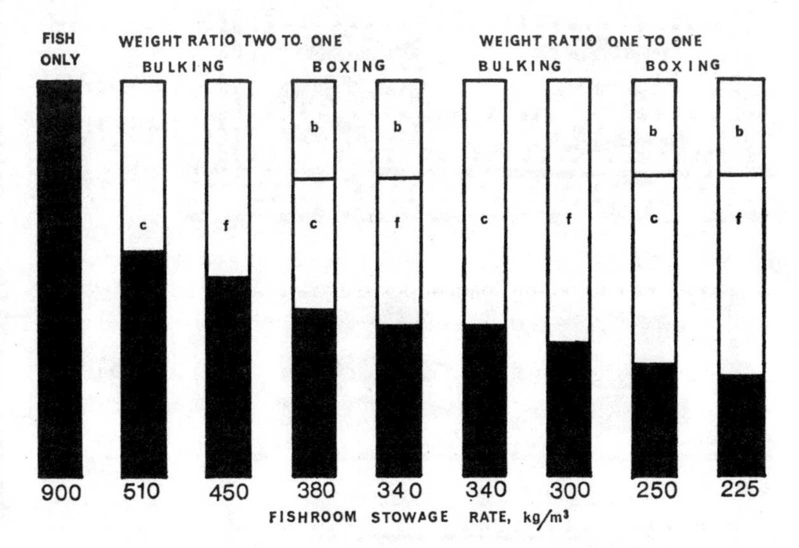

FIG 2.2 Typical stowage rates for wet fish

b: boxes c: crushed ice
f: flake ice

icing may be reduced somewhat. In some cases the fish is not refrigerated at all. High catching rate often implies a short voyage and hence a short period of storage; on the other hand spoilage can take place rapidly, especially at high temperature as indicated in Fig 1.1.

Methods of limited icing have been employed widely in fisheries ranging from near to distant waters. One method, shelfing, has been described. Fig 2.3 gives some idea of the many variations for fish iced in boxes. It is difficult to choose an exact order of preference because the results will depend partly on fishroom conditions, the amount of ice employed and the design of the box. Method (a) is the ideal method and method (e) is the worst of those shown because it does not aim to make use of the cooling effect of meltwater. In some fisheries the fisherman does not put ice directly on top of the top fish in a box. Either the top is not iced at all or a sheet of waxed or greaseproof paper may be placed directly on top of the fish as shown. This is done in order to protect the appearance of the fish for marketing, in the same way as in shelfing, with some sacrifice of eating quality. It is recommended to adopt the ideal method (a) with good mixing of fish and ice and using plenty of ice

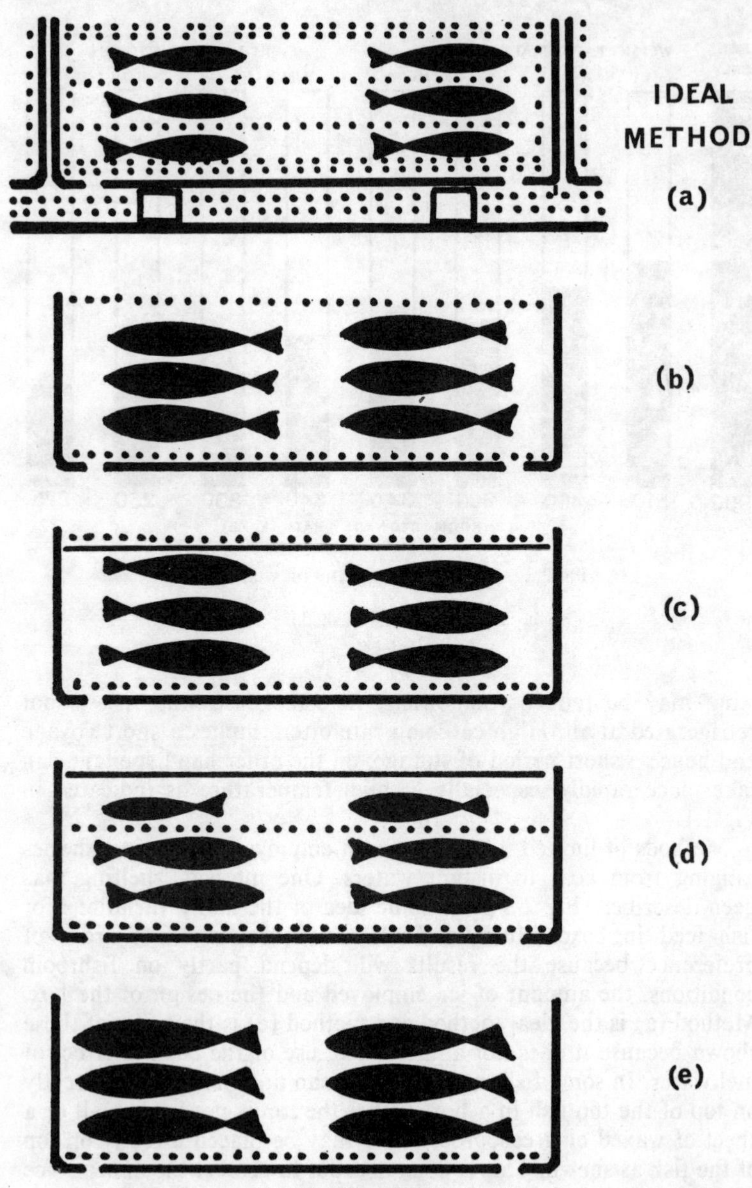

FIG 2.3 Methods of icing
.........layer of ice
——sheet of waxed paper

to allow for depletion, paying special attention when there are warm conditions in the fishroom. If necessary, ice should be placed around the outsides of the boxes in order to prevent the entry of heat.

It is to be expected that ambient conditions in the fishroom will have some influence on the rate of spoilage, especially where icing practice falls short of the ideal. There is only a limited amount of information on this point because few measurements have been made under commercial conditions. Of course heat gains into the fishroom should be held to a minimum in order to avoid rapid depletion of ice. This approach has led to the installation of refrigeration plant in many vessels, in order to provide cooling as a supplement to ice. Nevertheless as a rule when fish temperatures are too high the remedy is straightforward; more ice should be used in the right places.

SUPPLEMENTS TO ICE

Antibiotic ice

There have been a number of proposals for the introduction of measures other than cooling to inhibit bacterial action and thus extend the shelf life of iced fish. The addition of antibiotics and of mild acids are good examples. In some countries the use of antibiotics for fish preservation is not permitted on public health grounds. Traces of anbtibiotic will remain in the fish even after washing and cooking.

The antibiotics chlortetracycline and oxytetracycline have been included in ice at the time of manufacture, in concentrations of up to 5 parts per million and used commercially, in some cases for only the first part of the catch. Icing is carried out in the usual way but there is some indication that, with prolonged use of antibiotic, the patterns of population of spoilage bacteria in the fishroom are changed.

The extension of shelf life is not very great. Where there is a storage life of 14 days for cod in ordinary melting ice, for example, there will be a storage life of 15 to 17 days in antibiotic ice. There is no significant effect during the first 10 days of storage. Thus the rate of spoilage is retarded in only that part of the catch which has already passed out of the very fresh category. In fact this is often the case with anti-bacterial measures other than cooling.

Salt water ice

The attraction of ice made from clean sea water stems to some extent from the fact that its melting point is lower than that of freshwater ice due to the presence of salts. Therefore one would expect a reduction in the rate of spoilage with sea water ice. Reports on the storage of fish in sea water ice are not in complete agreement but without doubt there is little if any extension of storage life.

Freezing tends to separate the salt and water, the amount of separation depending on the rate of freezing. No appreciable separation is encountered in the flake ice machine, however, some of which have been built for the manufacture of ice from sea water. There is also separation of the salt and water on melting, so that over a period of several days the salt tends to drain out, leaving behind substantially freshwater ice. Thus, although the fish temperature may be lower initially, it gradually rises to the value obtained with freshwater ice.

Sea water ice can be used where freshwater is in short supply. The installation of an ice-making machine on board will not be feasible in the majority of cases where it is convenient to store ice in the fishroom. Flake ice machines have been used on board factory vessels which do not employ storage in ice as the main method of preservation but require some means of chilling the fish before and during processing and supply ice to catcher vessels.

The melting point of sea water ice is roughly $-2\,^{\circ}C$. Bearing in mind the elevation of the melting point on melting, there is no danger of damage to the fish by slow freezing, even with lean fish which begin to freeze at about $-1\,^{\circ}C$. The freezing point of freshwater or sea water can be reduced, however, by the addition of salt. Thus the fish can be chilled below the point where ice begins to form in the flesh. Bacterial spoilage can be retarded in this way but this will be of little significance over the first few days of storage. The fish will suffer damage through partial freezing and possibly unwanted salt contamination. Storage at temperatures slightly below the freezing point of the fish, superchilling, in the region of $-2\,^{\circ}C$ has been practised and is discussed in Chapter 5.

Gas storage

There can be an appreciable build up of carbon dioxide gas in the fishroom with ordinary storage in melting ice due to the action of spoilage bacteria, especially when the fishroom has been filled with fish and ice and closed tight for some time before discharge. It has been observed by some members of the distant water fishing industry that preservation is slightly better when the fishroom has been completely filled, perhaps due in part to the presence of carbon dioxide which inhibits the growth of spoilage bacteria.

Although the deliberate introduction of carbon dioxide gas can make a significant further reduction in the rate of bacterial spoilage, the method is not attractive. Apart from the practical difficulties involved in carrying the gas on board and maintaining the desired concentration of 30 per cent or so, the flesh of the fish tends to become bleached and soft. In experiments in which the oxygen content of the atmosphere was depleted by the introduction of nitrogen gas, there was no extension of shelf life.

Chapter

The Fishroom 3

CLEANLINESS

The fishroom must be kept in a reasonable state of cleanliness in order to keep spoilage bacteria and odours in check. It should be cleaned thoroughly once a week or, if the period must be longer due to the length of voyage, after the fish has been discharged at the end of the voyage. Unused ice should be discarded because it will carry a high bacterial load, unless it has been held at a temperature much below the freezing point.

Removable parts such as boxes and boards normally should be cleaned on shore. Sometimes they are cleaned at a central plant serving a port or area, where repairs and replacements can be made more easily. Since the bacterial load will have increased on unused wooden boxes and boards, they too should be cleaned.

The fishroom itself should be washed thoroughly, hosing down with clean mains water and using a stiff brush where necessary to remove fish slime, dirt, etc. The water pressure should be 150 kN/m^2 gauge (1·5 b) or more. It is preferable to use a detergent but care must be exercised in its choice to avoid tainting of the fish. It can be put into the water or applied directly. As a last step, the room should be hosed carefully with water so that no foreign material or detergent remains.

Drainage

The fishroom should have a good system of drainage. Shelves, boxes or whatever system is employed to support the fish and ice should be designed with drainage in mind. Contaminated meltwater from one compartment should not be drained through the fish and ice in another.

The drainage system can be a source of trouble if it is not maintained in clean condition. Dirty water lying in the system can create unpleasant odours which might be absorbed by the fish. Potential trouble points such as slush wells should be treated with a suitable disinfectant before stowage begins.

Fishroom Linings

The ideal lining for a fishroom would have the following properties.

- It would be watertight.
- It would have a hard and smooth surface which would not harbour bacteria and dirt.
- It would be easy to clean.
- It would be robust; able to withstand blows inflicted with pound boards, ice axes, etc.
- It would reflect light to a high degree, an important factor for good working conditions.
- It would not contaminate the fish.
- It would not give rise to corrosion.
- It would be light in weight.

Serious corrosion by electrolytic action can occur when dissimilar materials are in contact, especially in the presence of water and salts. The decay of metal and wood can be caused in this way. Contact can be broken by a material which is electrically non-conducting. The bottom or floor of the fishroom would be lined with a material having all the above properties with emphasis on robustness but it would not be slippery.

Linings for the sides and deckhead are commonly of wood or metal installed on the frames of the vessel so that there is a space between the linings and the hull. Wood absorbs moisture and is difficult to clean. It can be rendered waterproof, however, by a plastic coating but the better coatings are costly. Sometimes sheet metal is installed over a wooden lining. Half-hard aluminium alloy free of copper is a common type of lining. There is considerable interest in fibreglass reinforced plastic which closely approaches the ideal and is easily repaired in the event of damage, although its first cost usually is higher than the cost of a metal lining. Concrete linings have been employed in some cases but have few advantages and are not so easily cleaned and maintained. Wood and reinforced concrete, although they can harbour spoilage bacteria, and granolithic concrete are commonly used for the floor.

A lining is not always installed in a wooden vessel. The presence of linings in a wooden vessel tends to encourage wood rot. This can be countered by pre-treatment of the wood with a preservative to protect it against attack by fungi and marine borers, preferably by pressure impregnation, and by ensuring that the space between lining and hull is ventilated.

Heat Gains

Since heat in the fish will speed up spoilage, we should recognize the sources of heat. Icing practice and design of the fishroom then can be made more effective. The usual ways for heat to enter the fishroom are as follows:

- the fish
- bacterial action
- electric apparatus
- the fisherman
- air changes
- pipes, stanchions, etc.
- bottom, sides and deckhead
- bulkheads

The fish

The fish usually will be above 0 °C when they are stowed in the fishroom. They should be cooled to 0 °C as soon as possible after they are caught. This is one of the tasks performed by the ice in the fishroom, as explained in Chapter 2. The specific heat of the fish is about 4·0 kJ/kg degC.

Bacterial action

The ice also will have to cope with some heat generated by the action of spoilage bacteria on the fish. The amount of heat can be appreciable during the latter stages of spoilage when there are large numbers of bacteria present.

Electric apparatus

Apparatus which generates heat should not be placed in the fishroom unless it is necessary. Electric lights and motors usually are the only items in this category. Heat from lights will tend to increase fishroom temperature and thus indirectly can cause an increase in fish temperature if icing is inadequate. Good lighting is essential for good working conditions but it is worth remembering that the amount of power required for a given standard of lighting depends a great deal on the reflectivity of the surfaces of the fishroom. Highly reflective surfaces and light colours are best. In some cases it will be feasible to use fluorescent lighting for which the power required is less than half the power for incandescent lighting. Lights should not be left on unnecessarily.

The fisherman

Heat from the fisherman himself may be significant, especially in small fishrooms. The rate for one man will be approximately 0·30 kJ/s.

Air changes

In a fishroom properly designed for stowage in melting ice, the hatch through which the fish pass should be small. It is usually only large enough for a man to pass through and, in order to keep the rate of change of fishroom air at a low level, it should be open no longer than necessary. There should not be more than one opening at any time during storage because more than one opening may permit a great deal of air to circulate through the fishroom, in one opening and out another. Although large hatches are required for some modern methods of unloading the fishroom, especially where boxes or containers are used on large vessels, a small hatch for stowage at sea can be installed in the main hatch.

The rate of air change will depend on a number of factors including the size and location of the hatch but it is normally much higher than required for those working in the fishroom. The ventilation requirement of the fisherman working in the fishroom is about 10 dm^3/s per man and a hatch opening the size of a manhole normally supplies much more. There may be scope for hatches designed to reduce the rate of air change, particularly for small fishrooms where the influence of the hatch opening is relatively great.

Where there is a possibility of a build up of carbon dioxide in a fishroom which has been shut tight for a long period, it is essential to ensure that there is adequate ventilation for those who must work in the fishroom, usually when the catch is unloaded. In fact there is a risk of severe injury or death where fish, especially ungutted fish, is stored in bulk at high temperature without much ventilation. Oxygen levels will tend to fall as the level of carbon dioxide and other toxic gases increases. Dangerous gases may be released suddenly when the catch is disturbed and, since they are heavier than air, can remain in the bottom of the empty fishroom. Special precautions must be taken; sleeping on board in port may be inadvisable and close supervision must be exercised over fishroom operations. Expert advice should be sought if necessary, especially when ungutted fish are stored in bulk without adequate refrigeration, whether on ship or on shore.

The heat introduced by outside air can be considerable. If it is assumed that the fishroom is to be maintained at 0 °C, then air entering from outside will have to be cooled to 0 °C. Thus at a ventilation rate of 30 dm^3/s, outside air at 20 °C would impose a cooling load of up to 1·75 kJ/s depending on the humidity. Outside air at 10 °C would impose a load of up to 0·72 kJ/s, an appreciable load even for a large fishroom. The ventilation rate of 30 dm^3/s is lower than those encountered in practice, where in fact excessive rates of ventilation make it impossible to maintain low temperatures, especially in the smaller fishrooms. A ventilation rate of 30 dm^3/s would provide 2 air changes per h in a fishroom of 54 m^3 and $\frac{1}{2}$ air change per h in a fishroom of 216 m^3.

Pipes, stanchions, etc.

Since metal has such a low resistance to heat flow, a metal hull will be at the sea temperature and heat will be carried readily into the fishroom along any metal parts, pipes, conduits, etc., in contact with the hull. In the same way heat can be carried into the fishroom from the engine room and other parts of the vessel. Supports and fixings for pipes, linings and the like are also potential trouble points. These difficulties can be reduced by judicious use of materials with a relatively high resistance to heat flow such as plastic and wood.

Slushwells can be a serious source of heat gain to the fishroom but it is possible to reduce heat gain by good design. The tank can be separated from the hull by waterproof insulation and pipe connections or entire runs of pipe can be made of plastic.

Fishroom stanchions are often connected to the hull in order to contribute to the structural strength of the vessel. It is advisable to interpose sections of sufficient strength but with a higher resistance to heat flow than metal. Again, wood and reinforced plastic can be used.

Bottom, sides and deckhead

Heat from the outside, entering the fishroom through the surfaces, can cause an increase in fishroom temperature and thus indirectly can cause spoilage if the fish is not adequately protected by ice. With most methods of stowage, notably bulking, it is essential that the fish stowed near the surfaces of the fishroom are protected by a layer of ice from heat directly from the surfaces. Some ice should be used for this purpose even in insulated fishrooms although the amount of heat and hence the amount of ice required can be reduced by insulation.

Bulkheads

Bulkheads must be considered in much the same way as the other surfaces. Often, however, the temperature difference across a bulkhead is high compared with the differences at other surfaces. Hence the heat flow across bulkheads sometimes tends to be high, particularly with engine room bulkheads.

Chapter

Fishroom Insulation 4

INSULATION

THERMAL insulation is a material having a high resistance to heat flow. It can be installed between the hull and the lining and over other surfaces, therefore, to retard the flow of heat into the fishroom and so reduce the amount of refrigeration required.

Provided sufficient ice is used with proper stowage, the result will be satisfactory without insulation. In practice, however, even in some cases where the voyages are of short duration in cold climates, it is necessary to insulate the fishroom

- to reduce fishroom temperature,
- to conserve refrigeration,
- to help prevent spoilage of fish stowed near the surfaces where heat enters.

It is important to remember that ice should be used against all surfaces to protect the fish from incoming heat and prevent bilgy spoilage, even when these areas are insulated.

Presence of water

There are many insulation materials which might be used for fishing vessels where melting ice is employed but few of them are highly satisfactory, mainly because of moisture problems. The suitability of an insulation and the best method of insulation will depend on the type of hull and type of lining. Often the lining is far from watertight and there may be some leakage of water through the hull, so that the insulation will be attacked by water and perhaps by rot. Superimposed on this problem of liquid water seepage directly into the insulation, there is a likelihood that water vapour will condense in the insulation, on the inside of the hull, or on the inside of the lining, depending on the temperatures. Usually there are considerable variations in temperature which can give rise to changes in the movement of water vapour in the space between lining and hull. The outside temperature will be higher than the fishroom temperature during most voyages but it may be lower during winter conditions or when the vessel is idle.

The presence of water in the insulation can diminish or destroy its value. Thus insulation materials which are impervious to water are most attractive. Even when the insulation material itself has a low permeability, however, water or water vapour may pass through joints or gaps in the insulation.

Methods and materials

The type of ship construction will influence the question of insulation. The hull of the fishing vessel commonly will be of wood or steel. There is considerable interest, however, in a number of other materials including aluminium, fibreglass reinforced plastic and concrete. Wooden vessels are generally smaller with smaller fishrooms than steel vessels and the voyages are usually of shorter duration.

The arrangement of linings and insulation and the method of installation will depend on the materials employed and the design of the vessel. The insulation must be protected against the ingress of water. Wood grounds for the fixing of insulation should be treated against rot. Precautions against rot will be a major feature of installations in wooden ships. The insulation must not be installed with gaps in the joints because bad joints will reduce its effectiveness and permit the passage of water and water vapour. Advanced methods and materials can give a completely watertight assembly of hull, lining and insulation. There has been considerable success with plastic insulation foamed in place, completely filling the space between steel hull and linings of fibreglass reinforced plastic. This method also would be suitable for vessels with metal linings and hulls of aluminium and reinforced plastic. Flammable materials should not be used in an insulation system, in order to keep fire risk at a minimum. Generally the plastic materials do not burn readily but there is danger from toxic fumes from burning plastics. Some regulations demand fireproof linings and specify certain permitted insulation materials.

Insulation for the floor of the fishroom must have an adequate compressive strength, usually not less than $20\,t/m^2$. In larger vessels it may be applied over storage tanks for water and fuel. The passage for the main shaft, if it runs under the fishroom floor, should be covered by insulation or at least a cover of waterproof wood of 5 cm minimum thickness.

Hatch covers should be insulated and lined.

Particular attention must be given to insulation of bulkheads, especially bulkheads separating engine room from fishroom. The engine room is normally at high temperature, typically 30°C, and there have been many instances where it has not been sufficiently insulated from the fishroom.

Cork, mineral wool, glass wool, expanded ebonite and expanded plastics and some other materials have been employed as insulation in fishing vessels. Expanded materials with closed cell structure, such as ebonite, polyurethane and polyvinylchloride which have low permeability, are usually more suitable for fishrooms. If this type of insulation is used under a wooden lining or another type of porous lining, the two should be separated by an air space ventilated to the fishroom, otherwise the lining may become unduly wet and a source of smells and rot. There should be an air space, ventilated to outside atmosphere, between the hull and insulation in wooden vessels. Expanded plastics of the open cell type, cork and wool insulations are at a disadvantage because they have high permeability. Systems which employ these materials must include provisions to keep them dry or they will not remain effective for a reasonable period. Often the insulation is installed behind a watertight lining with a ventilated air space between the insulation and hull.

The use of linings without insulation, particularly wooden linings, is common in large steel vessels in arctic waters. The resistance to heat flow into the fishroom, largely provided by the air space between hull and lining, is only moderate. There can be serious spoilage if the fish are placed against the lining. With reasonable cleanliness and sufficient ice against the lining, satisfactory results have been obtained but there is no doubt that this type of fishroom is far from ideal, especially for warmer climates.

Many smaller wooden vessels have only the hull with no linings or insulation. Bearing in mind the relatively great influence of the hatch and the higher surface to volume ratio of the fishroom, fishroom temperature will tend to be higher and less influenced by linings and insulation in the smaller vessel. In cases where the catch is stored in ice in boxes with no ice at the surfaces of the fishroom, linings have little effect, typically giving a reduction of 1 degC in a fishroom of 50 m^3. Insulation also will have a limited effect under these circumstances. Hence the interest in insulated boxes and systems of mechanical refrigeration to reduce the rate of depletion of the ice and improve storage conditions. It is important to remember, however, that the desired fish temperature of 0°C can be maintained by ice alone if enough ice is employed inside and outside the boxes.

Properties

Steel has a low resistance to heat flow, dry wood has moderate resistance and cork has a high resistance as indicated in Table 1.1. An air space has an appreciable resistance, somewhat dependent on the amount of air circulation in the space and the reflective properties of the boundary surfaces. Still air and highly reflective surfaces will give the highest resistance. The resistance column in Table 1.1 illustrates the point that metal has a very low resistance to heat flow. All metal frames, pipes, conduits, etc., connected to a steel hull can conduct heat rapidly from the hull to the fishroom. The fish must be protected from this heat as well as heat from other sources. Also, the relatively high resistance of insulation materials is illustrated. It should be noted that fish itself has a high resistance.

Some values of thermal resistance of insulation materials are given in Table 4.1. They give an indication of some of the insulation materials of interest. There are other materials of interest, of course, and the properties of those listed can vary a great deal depending on the density, method of manufacture and other factors. In some cases properties other than those listed, such as flexural strength and thermal coefficient of expansion, will have a bearing on the choice of material. The insulation can be in any of a number of forms; loose, board, blanket and foamed in place plastic insulation.

Differences in ability to withstand rot and chemical attack often are not great enough to have a major bearing on the choice of insulation material. Cork is the least attractive of the materials listed in the table from this point of view. Only glass wool and mineral wool are virtually unaffected by fire and chemicals but their lack of rigidity and strength and high permeability are disadvantages. The materials listed with low permeability, all expanded with closed cell structures, are usually more expensive than the others. On the whole they are quite stable, largely because they are impervious to moisture, but adhesives to be used with them must be chosen carefully if fixing is to be secure.

How much insulation?

The amount of insulation required depends on a number of factors; climate, vessel construction, method of stowage, length of voyage, etc. There is a strong case for at least 5 cm of insulation in all but the coldest climates, giving an overall value of thermal resistance greater than 1·4 m^2 degC/W. The choice of temperature difference for design purposes will depend on the climate and conditions. Solar radiation will be an important factor, especially in latitudes 0 to 50 degrees. The surface temperature of exposed deck and hull can rise to 20 degC above the ambient temperature, depending on the reflective properties of the surface. Light coloured surfaces will absorb less radiation than dark surfaces. The surface temperature and hence heat gain through the deck sometimes can be reduced by hosing it with water.

Where fishroom capacity is an important factor it may be better to install a given thickness of a more expensive insulation than a greater thickness of a cheaper insulation with the same resistance. In a fishroom of 60 m^3 with a surface area of 100 m^2, for instance, 5 cm of insulation would occupy a volume of 5 m^3, which is more than 8 per cent of the fishroom capacity. The actual reduction in fishroom capacity may be less than 5 m^3, depending on the construction.

As in the example below, the insulation may give a considerable reduction in the heat gain to the fishroom. Remembering that the heat entering through the fishroom lining should be absorbed by melting ice, the saving in ice can compensate for the space taken up by the insulation. Where ice is placed against the lining to absorb heat directly, the melting rate in the example will be reduced by about 1 kg/m^2 h, or about 50 mm of flake ice per day.

Although this question of conservation of ice may be important, it will take second place to the problem of spoilage where fish is stowed near areas of heat gain, including dehydration of the fish in severe cases. As the trend toward grading for quality increases, largely by the introduction of easier and better methods of quality assessment on landing, the fisherman will be forced to pay even closer attention to proper handling and stowage on board.

High fishroom temperatures, particularly above 10°C, often lead to poor results largely because the ice is depleted too rapidly, leaving the fish inadequately protected. Where the fish is not completely surrounded by ice and partly exposed to the fishroom conditions, spoilage will be encouraged, especially at high fishroom temperature. There is not much information on the influence of fishroom conditions on fish temperature with the less effective icing techniques. With one method of icing in a wooden box, shown in Fig 2.3(c), a record of fish temperature over a period of 33 h showed an initial reduction to 1·7°C in about 4 h and then an increase to 4·5°C because of depletion of ice. The resulting quality was equivalent to the quality of fish stored at 0°C in ice for 60 h, which represents a loss of shelf life of more than one day. The

Table 4.1 Properties of Insulation

Insulation material	*Density, kg/m³*	*Resistance for 5 cm thickness, m² degC/W*	*Compressive strength, t/m²*	*Permeability to water vapour*
corkboard	150	1·40	50	high
	200	1·00	70	high
glass wool	70	1·40	nil	high
mineral wool	70	1·40	nil	high
expanded polystyrene				
open cell	16	1·75	6	high
closed cell	30	1·50	25	low
expanded polyurethane	25	1·40	12	low
expanded polyvinylchloride	20	1·60	10	low
	40	1·60	30	low
expanded ebonite	70	1·75	30	low
	200	1·25	120	low

fishroom was uninsulated and at 11 °C. Probably fish temperature would have been even higher if metal boxes had been used because the wooden boxes had some resistance to flow of heat from the fishroom. One approach to this problem is to employ boxes made of insulation material as discussed in Chapter 2. In any event the emphasis should be on the use of plenty of ice, properly disposed to absorb heat which might otherwise reach the fish.

The engine room bulkhead is an important area. Too often it is poorly insulated or not insulated at all. Fish stowed near such a bulkhead for a few hours can suffer a marked amount of drying if they are not completely protected by ice. In a fishroom of 60 m^3 the heat flow through an uninsulated bulkhead might be higher than 0·50 kJ/s, enough to raise the temperature of an unlined wooden fishroom by about 1 degC. The increase in temperature will tend to be greater in a lined or insulated fishroom.

Example

According to the considerations outlined in Chapter 1, the relationship between heat flow rate through a structure such as a bulkhead, ship's side, etc., temperature, area and resistance can be expressed as follows:

$$H/A = D/R$$

where H = heat flow rate, W
A = area of structure, m^2
D = temperature difference, degC
R = thermal resistance, m^2 degC/W

In the following example, heat gains are estimated for an uninsulated and an insulated fishroom structure shown in Fig 4.1. These gains will form part of the total cooling load. There is water on the outside and melting ice at the inside surface, practically equivalent to water. The temperature difference between outside and inside is 20 degC and the area is 50 m^2. The thermal resistance of the insulation is 1·75 m^2 degC/W.

The resistances of the components, found in Table 1.1, must be added to obtain the total resistance of the structure shown in Fig 4.1(a).

Item or component	*Resistance, m^2 degC/W*
outside surface	negligible
steel hull	negligible
air space	0·17
wet wood lining	0·09

inside surface	negligible
total	0·26

An exact calculation is difficult. The steel frames of the vessel, which cross the air space, will have virtually no resistance of course, but the wooden grounds have about the same resistance as the air space. The resistance would be slightly lower if the lining were fastened directly in contact with the frames. The resistance of the air space will be more or less correct if it is not ventilated, even if the width is somewhat greater than 10 cm. Ventilation easily could reduce the total resistance to less than 0·20 m^2 degC/W. On the other hand, with air at the inside surface instead of melting ice, the value would be more than 0·30 m^2 degC/W.

The resistance of the structure shown in Fig 4.1(b) is estimated in the same way.

Item or component	*Resistance, m^2 degC/W*
outside surface	negligible
steel hull	negligible
air space	0·17
insulation	1·75
sheet metal lining	negligible
inside surface	negligible
total	1·92

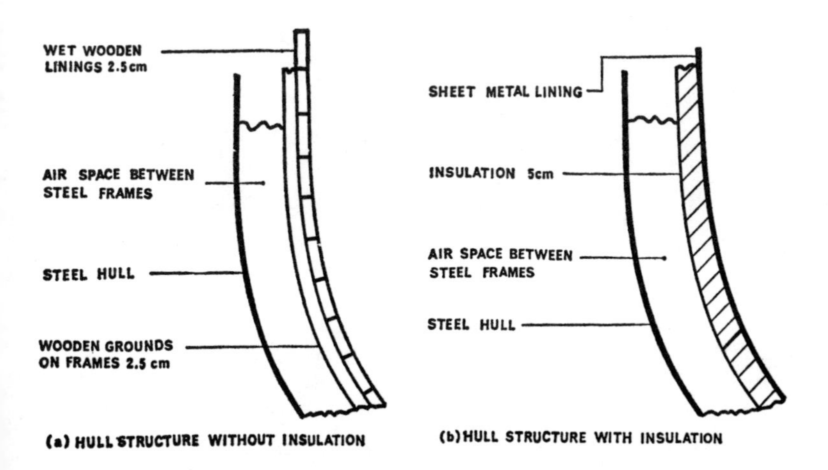

FIG 4.1 Fishroom insulation

It is seen that the resistance of the structure can be increased greatly by the addition of 5 cm of insulation. The actual value may be slightly lower than the calculated value because of fasteners through the insulation. Also, ventilation of the air space will reduce the resistance. It is common practice, therefore, where there is a generous and effective amount of insulation, to estimate the total resistance and heat gain based only on the insulation, ignoring the other components. Thus in this case the total resistance would be estimated at 1·75 m^2 degC/W. Doubling the insulation thickness would give a value of 3·50 m^2 degC/W.

The estimated heat gains through the structures 'a' and 'b' are calculated using the revised values of total resistance.

$$\begin{aligned} R_a &= 0{\cdot}20 \text{ m}^2 \text{ degC/W} \\ H_a/A &= D/R_a \\ &= 20/0{\cdot}20 \\ &= 100 \text{ W/m}^2 \\ H_a &= A \times 100 \\ &= 50 \times 100 \\ &= 5{,}000 \text{ W} \end{aligned}$$

$$\begin{aligned} R_b &= 1{\cdot}75 \text{ m}^2 \text{ degC/W} \\ H_b/A &= D/R_b \\ &= 20/1{\cdot}75 \\ &= 11{\cdot}4 \text{ W/m}^2 \\ H_b &= A \times 11{\cdot}4 \\ &= 50 \times 11{\cdot}4 \\ &= 570 \text{ W} \end{aligned}$$

The difference of 88·6 W/m^2 in this case corresponds to a melting rate of ice of 1·0 kg/m^2 h.

Chapter

Mechanical Refrigeration with Ice 5

MECHANICAL REFRIGERATION

IT has been emphasized in Chapters 2 and 3 that the catch can be properly chilled by ice alone or by ice with the aid of insulation. Why then should mechanical refrigeration be used? Broadly, the reasons are the same as those put forward for the use of insulation in the fishroom:

- to reduce fishroom temperature

- to conserve ice

When sufficient ice is used and when the icing technique is adequate, which unfortunately do not always occur, the fish will be protected from warm fishroom conditions. Low fishroom temperatures will help reduce melting rates, however, and the mechanical system can help prevent spoilage of fish stowed near the surfaces by intercepting heat gains. The introduction of mechanical refrigeration, particularly in a warm climate, can enable a reduction in the amount of ice required with proper icing technique. In an extreme case, it might mean a change from an overall fish to ice ratio of 1:1 to a ratio of 2:1. The reduction will tend to be greatest in the small fishroom with high heat gain per unit volume. Thus the work of stowage may be made easier and, where the size of the fishroom places a limitation on the duration of the voyage, the amount of fish landed may be increased. The fact that mechanical refrigeration can eliminate variations in icing technique from one season to another also will be an advantage in itself, notably where the fish are boxed at sea.

The use of mechanical refrigeration as a supplement to melting ice should be considered carefully in the light of climate, size of vessel, type of fishery and market, length of voyage, cost of ice and possibly some other factors. Often ice and insulation can do the job as well or better. When mechanical refrigeration is used, insulation is essential.

PIPE GRIDS

Probably the most common method of mechanical refrigeration of the fishroom is the refrigerated pipe grid. The grid is normally installed on the underside of the deck, the deckhead, perhaps with some additional grid on bulkheads or sides of the fishroom. Usually the pipe is of galvanized steel which is robust but sometimes coils of hard copper are used because of the weight advantage. A copper grid will have about one-quarter the weight of a steel grid.

Operation

If bulking is the method of stowage, ice placed against the sides, bottom and bulkheads will have to absorb the heat gains through these surfaces in the usual way, otherwise the heat will reach fish stowed near the surfaces. Deckhead grids will be able to cope with heat gained through exposed fishroom surfaces, especially the deckhead, and some of the heat introduced with outside air. The fish will be cooled by the melting ice, although the rate of melting may be reduced by the use of grids.

With a system of shelfing where there is space for natural convection of air, the grids will have more effect because the fish are partly exposed to fishroom air and larger areas of the surfaces of the fishroom will be exposed. Even so, it should be recognized that shelfing is inferior to proper bulking and boxing in ice. The cooling of the fish itself is best left to the melting ice.

With boxing there are various possibilities. Close stacking of the boxes with liberal use of ice will give practically the same situation as bulking. A more open type of stowage will allow the grid to exert more influence as with shelfing.

In some vessels more than one method of stowage is used on the same voyage. For example, the latter part of the catch may be shelfed after the rest of the catch has been bulked. Thus in larger vessels it may be desirable to divide the grid into two or more separate circuits with separate control. Cooling by mechanical refrigeration can be more or less confined to the area where shelfing is employed, usually forward in arctic trawlers.

In cold ambient conditions it is sometimes better to shut off the grid during stowage because heat gains and fishroom temperature will be low. In any case the mechanical system can be of benefit on the outward voyage. If the fishroom is cooled and held at 0°C or slightly below, melting of the ice will be reduced or prevented and it will remain crisp and easily handled. Also, the mass of the fishroom will be cooled before stowage begins, a job which otherwise might not be done by the ice. Development of bacteria on boards, fishroom surfaces, etc., and (if the temperature is appreciably below 0 °C) development of bacteria in the ice will be retarded.

Cooling capacity

It is difficult to estimate the cooling load but the information given in Chapters 3 and 4 will help to give some idea of the load. It is even more difficult to calculate the required capacity of a mechanical refrigeration system where it is to supplement the ice. Climate will be a major factor. Experience in various fisheries, however, has provided some guidance. For a fishroom temperature of 0 °C or slightly above in tropical waters, a small vessel with a properly designed and insulated fishroom of, say, 30 m^3 will have a system with a cooling capacity of about 100 W/m^3 or 3 kJ/s and a refrigerant compressor with an electrical rating of 1 kW. A larger vessel with a fishroom of, say, 180 m^3 will have a system with a cooling capacity of about 50 W/m^3 or 9 kJ/s and a compressor with an electrical rating of 3 kW. The corresponding capacities for arctic fisheries are about 30 per cent lower.

As a guide, the minimum required surface area of pipe grid installed in the fishroom is 1·0 dm^2/W and 1·5 dm^2/W is recommended. In the examples given above, this would mean a total grid area of at least 30 m^2 in the small vessel and 90 m^2 in the large vessel. The pipe diameter should be 2 to 5 cm. The grid normally is operated at a temperature of -3 to -5 °C. Much of the cooling will be by condensation of moisture in the form of frost on the grid. The two greatest sources of the moisture are the ice and air introduced from outside.

Defrost

A rapid defrost system for the grids is sometimes useful. Unless there is a defrost system it may be necessary to shut off the refrigeration some hours, perhaps as much as 48 h, before discharge of the catch in order to allow the frost which accumulates on the grid to melt and the grid to become dry. Dripping water from the grids can worsen conditions for persons unloading the catch from the fishroom. Defrost methods are described in Chapter 7. Usually a hot gas defrost system is employed for grids. In the larger installations it is feasible to divide the grid into four or more separate circuits, enabling a rapid and effective defrost of one section at a time.

Temperature control

The most difficult question in the operation of the pipe grid is that of temperature control. There is a danger that fish stowed near the grid will be subject to freezing conditions and possibly excessive dehydration. A thermostat must be located with care because the heat gains to the fishroom may vary from one part to another and the method of stowage may prevent the natural circulation of air. The store of ice will also exert an influence. The sensing element of the thermostat should not be placed near a warm surface, near an open hatch or too near the grid. It should be in contact with fishroom air only, not with ice or fish. An indicating or recording thermometer, preferably independent of the thermostat, with a sensing element near the thermostat element and with remote reading where necessary, should act as a check on the thermostat. It is good policy to check temperatures periodically here and there in the fishroom, particularly large fishrooms which should be equipped with two or more permanently installed thermometers with remote reading. Sensing elements and leads should be protected. The sensing elements are usually shielded by a perforated metal sleeve which protects against knocks, etc., but permits sufficiently quick response to changes in temperature. A thorough temperature survey under operating conditions will enable those responsible for stowage of the catch to become familiar with the fishroom and thus carry out their work more intelligently.

As an alternative to thermostatic control of the plant, manual control with the aid of thermometers can be used. Thermostatic control is preferred under most circumstances. If there are ideal conditions of temperature distribution and control, the thermostat can be set at 0°C, just above the freezing point of the fish. Usually, however, the setting is between 1 and 2°C in order to avoid the danger of freezing some of the fish and to allow the ice to melt.

Forced Convection Cooling

Cooling by forced convection, with a fan for air recirculation, a compact cooling coil in a convenient location and possibly ducts for distribution of supply air and return air, has been used in a limited number of cases. In one system for shelfed fish, air is discharged through a number of openings at both sides of the fishroom between shelves and returned through a common duct along the centreline of the room to the cooler at one end. The branch ducts for supply air are installed with the insulation between frames. Such a system gives more flexibility and better control of temperature than the deckhead grid but it is not worthwhile for bulked fish.

Boxing

Forced convection can be employed with boxing in much the same way as in cargo vessels carrying fruit in cartons. The boxes are arranged so that air can circulate around them, achieving an even temperature throughout the fishroom and absorbing the various heat gains. This method has been used with aluminium boxes which, although they have a number of advantages, have a low resistance to heat flow. The direction of air flow is normally from bottom to top of the open stack of boxes. A uniform space between boxes can be made by spacers or by projections at the sides of the boxes, incorporated at the time of manufacture. The size of space required is roughly 1 cm. Such techniques have not been applied widely; the preferred alternative is to employ ice outside as well as inside the boxes, particularly at the surfaces of the fishroom.

Heat gains

Heat gains are a potential cause of dehydration, causing weight loss and adversely affecting the appearance of the fish, as well as a potential cause of increased bacterial action. A rate of 1 kJ/s corresponds to a rate of evaporation of more than $1\frac{1}{2}$ kg/h. Fortunately this potential is not fully realized, largely because of the protection afforded by the ice. Humidity is also introduced with outside air of course, but adds to the cooling load. It is important that the cooling system be designed and operated so as to avoid drying of the fish. The fish should not be subjected to rapidly circulating air which can encourage evaporation. In any properly designed system a high proportion of the refrigeration will be directed to the surfaces of the fishroom in order to cope with the major heat gains at these areas.

Fan heat is a potential source of drying. It is imperative, therefore, that this heat always is removed by the cooler. For this reason the fan should discharge through the cooler, not directly into the fishroom.

Cooling capacity

The cooling capacity required in a forced convection system will be the same as in the pipe grid system with natural convection except that allowance for the heat generated by the fan will impose an additional cooling load. The capacity of the refrigeration plant will have to be increased accordingly. As a rough guide, the fan power will be equal to the compressor power and constitute $\frac{3}{8}$ of the gross cooling load but the exact values will depend on the characteristics of the system. Typical fan capacities for the small and large fishrooms mentioned above would be 2·5 m^3/s and 7·5 m^3/s, based on a temperature rise across the fishroom (difference between air entering the fan and leaving the cooler) of 1 degC at the net cooling capacities of 3 kJ/s and 9 kJ/s.

The conditions of operation of heat exchangers are along much the same lines as those for the air blast freezer in Chapter 8, except that the refrigeration load for the freezer is relatively high and the temperature is lower.

Defrost

As in the grid system, frost will accumulate on the cooler but, unlike the grid which is effective even with an appreciable deposit of frost, it should not be allowed to build up too far, otherwise heat transfer and air flow will be impaired and defrosting may be made more difficult. The most suitable type and frequency of defrost will depend on circumstances. The fan should be turned off during defrost. There are several possibilities; hot gas, electric defrost and water spray. Defrost also is discussed in Chapter 7 and for the air blast freezer in Chapter 8. Control of the defrost can be exercised manually or automatically through a time switch. The hot gas defrost is suitable where there are several coolers in the one system.

Temperature control

Temperature control is relatively easy and the choice of location for the thermostat is straightforward. The thermostat should be located in the supply air. As a check, a record of supply air, return air and possibly fishroom temperatures should be made using an indicating thermometer or an automatic recorder independent of the thermostat. The supply air should be maintained at 1 °C or slightly higher, just above the freezing point of the fish.

THE JACKETED FISHROOM

In any cold store or chill room, including the fishroom, the heat which passes through the insulation can be intercepted by a refrigerated air space between the insulation and the lining, a jacket as shown in Fig 5.1. The jacket space is cooled by a fan-cooler unit located at some convenient point.

The jacketed store has been advocated mainly because the amount of heat entering the storage space, hence the potential for dehydration and spoilage, is reduced. Another advantage of the

FIG 5.1 Jacketed fishroom

method, however, is that the amount of insulation and the amount of mechanical refrigeration can be balanced to give minimum cost for the installation, dependent only on the temperature to be maintained inside the store. Thus the most economic thickness of insulation might be less than in the ordinary store, where some heat enters through the insulation.

A number of jacketed stores of various sizes have been built on land. The reported costs have varied from slightly lower to about 25 per cent higher than a conventional store. There is not necessarily a loss of storage space with jacketed construction because, according to normal good practice, there should be an air gap for the removal of heat between the goods and the inside surface of the store anyway. It is the usual practice to seal the jacket lining so that there is no leakage of air through the lining into the cold store or vice versa. Also the warm side (outside) of the insulation is sealed against the ingress of air and water vapour. Thus the cooler can be operated for long periods without a defrost.

Because the ratio of surface area to volume is higher for smaller stores, depending somewhat on the shape of the store, the heat gains through the surfaces will be proportionately higher. In view of this, the jacketed design is attractive for the comparatively small store such as the fishroom.

Fishroom

The jacketed fishroom has brought no significant improvement in temperate climates and it is doubtful whether it would be worthwhile even in tropical climates where the heat gains tend to be greater.

In many respects, because of the shape of the fishroom and the presence of water, installation of the jacketed system is more difficult on board than on shore. Fans and coolers should be accessible for maintenance through a removable section of lining or a section of insulation. It may be feasible to install it outside the jacket, with ducts through the insulation. The presence of water in the jacket space will make regular defrosting of the cooler necessary. The insulation preferably should be impervious to water, as in the conventional system.

The temperature of the jacket space must be controlled within very close limits at about 0°C in order to avoid freezing of the fish on the one hand and heating on the other. This is especially important where ice is placed directly against the fishroom lining, as in bulking, otherwise no benefit will be gained from a reduction in the thickness of the ice layer at the lining. Reduced losses due to dehydration, normally the main feature of a jacketed store ashore, should not be an issue here because there will be no dehydration of the fish properly iced.

Semi-jacket system

Another approach for the fishroom is the semi-jacket system in which air from the cooler is circulated between linings and insulation, but with the jacket open at the top and with no jacket at the deck-head. Thus the cooler is able to cool the fishroom directly, coping with heat gains from air changes, lights, etc., as well as the heat gains through the insulation. Such a system is feasible with grids with gravity circulation of air. The usual pound system can be modified slightly to provide the necessary linings where bulking is the method of stowage. Where the fish are stowed in boxes, a ventilated space at the bottom, sides and deckhead is easily arranged, even without linings or partitions.

Conduction jacket

What is effectively a jacket system can be made by the installation of cooling pipes or similar heat exchanger, sometimes in a layer of cement, between the lining and insulation with no air space. The assembly should be watertight because moisture will tend to migrate to the cold pipes in the same way as it does with the normal exposed grid. The fishroom lining and the outside of the insulation, which should be of the impervious type, should be sealed against the entry of water and water vapour. Alternatively, the pipes may be embedded in an impervious insulation. This type of cooling system has a high first cost but has been used in superchilling.

SUPERCHILLING

Superchilling is the cooling of the fish to a temperature between −1 and −3°C, just below the temperature of melting ice, where some of the water in the flesh will be frozen. Since the action of spoilage bacteria is by far the main cause of deterioration in iced fish at about 0°C and bearing in mind that bacterial and enzymic activities are much dependent on temperature, virtually ceasing at −5°C, one might expect to gain an extension in storage life by superchilling. To a limited extent this is so but partial freezing has some disadvantages.

The amount of refrigeration must be controlled carefully. Cooling will be slow unless the fish are subjected to temperatures much below the superchill range. Typically, the cooling down period will last for seven days. Faster cooling, entailing temperatures in the flesh well below −3°C and resulting in some denaturation of the protein, has been used in shore processing, particularly for fillets and other products prior to distribution. With suitable wrapping, the fish will thaw only slowly and thus remain at about 0°C for many hours without the need for additional refrigeration. With slow cooling there will be little denaturation but the formation of large ice crystals will cause some physical damage to the flesh. This disruption due to ice formation appears as a roughness at the cut surfaces such as the surface of a fillet. It has virtually no effect on the eating texture. Some of the more delicate species of fish with relatively soft flesh, such as hake, are considered to be less suitable for superchilling because they are less able to withstand this damage.

Partial freezing also may complicate the handling and processing after landing. When fish is stored in ice in this way, the meltwater refreezes at the reduced temperature and the whole mass of fish and ice tends to congeal. This can create difficulties in the discharge from the vessel and subsequent handling. The fact that the fish itself requires thawing also can create difficulties, particularly if filleting or other processing operations are to be carried out before distribution. It may be necessary to have special apparatus and facilities for thawing.

On the other hand, the refrigeration which has been stored in the fish in the form of ice in the flesh can give a certain amount of protection after landing. This might be useful in transport after landing. Heat from the surroundings otherwise might cause an appreciable increase in temperature and hence increase the rate of spoilage.

In this chapter the discussion will be confined to superchilling systems in which demersal fish are stowed in freshwater ice in more or less the usual way. Superchilling can be carried out in other ways; salt water ice is discussed in Chapter 2 and refrigerated sea water is discussed in Chapter 6.

Quality factors

The method has been used to only a limited extent by trawlers in warmer waters and hardly at all in the arctic fishery. This limited experience with demersal species has shown that the physical damage to the flesh due to ice formation does not completely offset the advantage of reduced bacterial action. The amount of physical damage is dependent, however, on the amount of freezing which has taken place. Cod, which begins to freeze at about $-1\,^{\circ}C$, is one-half frozen at $-2\,^{\circ}C$ and nearly three-quarters frozen at $-3\,^{\circ}C$ as shown in Fig 1.2. At $-3\,^{\circ}C$ the limit of acceptability based on eating quality is reached at about 35 days but the physical damage is appreciable and, although the fish may be suitable for some products where this change in the appearance of the flesh is not of great importance, the fish will be unsuitable for filleting and smoking. At $-2\,^{\circ}C$ the limit is about 26 days and the amount of damage is within acceptable limits for most purposes, showing as a slight roughness at the surface of the fillet. With the fish one-quarter frozen at about $-1\,^{\circ}C$ damage is slight and the limit of acceptability based on eating quality is reached after 20 days, as opposed to the normal 14 days for cod held at $0\,^{\circ}C$ in melting ice.

A typical result is shown in Fig 5.2. Although the eating quality of the superchilled fish is higher over the entire storage period, the ordinary iced fish will be preferred over the first 10 to 12 days because of the damage due to freezing in the superchilled fish. There is little point, therefore, in the use of superchill temperatures on board where the period of storage is less than 12 days. Thus superchilling cannot be looked upon as a means of producing fish of very good quality. In this respect there is some parallel between superchilling and the use of antibiotic ice which has little effect over the first 12 days of storage. Storage in ordinary melting ice is preferred for the shorter periods. Proper quick freezing and cold storage will give better quality with a much longer storage life.

Thawed superchilled fish should not be used for the production of frozen fish products. The changes in texture are too severe. Since

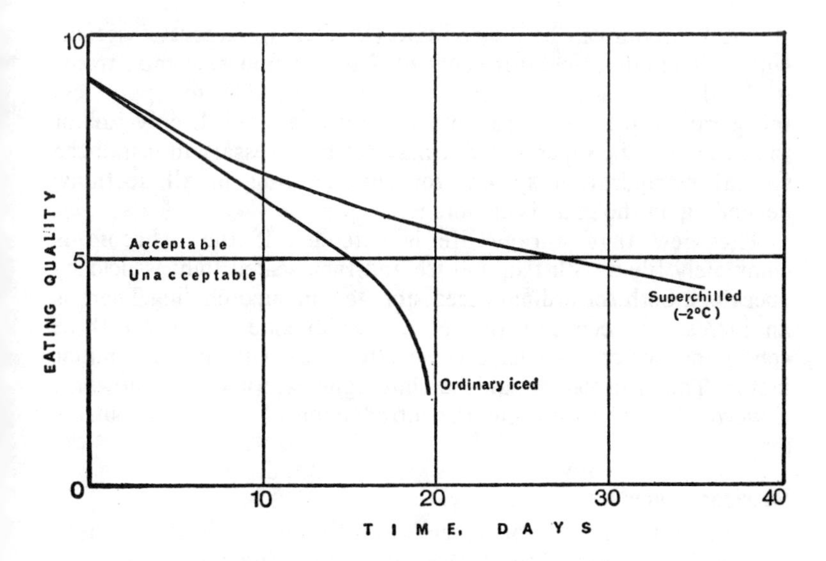

FIG 5.2 Storage life of superchilled cod

the changes in texture are the main factors affecting the quality of fish in quick freezing and cold storage, there is little point in the freezing of material which has already undergone an appreciable change. The only possible exception is where material is frozen for later manufacture of fish cakes or similar products because the flesh is broken down in any case and perhaps some increase in drip loss can be tolerated. Fish held in ice for more than 12 days will not be suitable for freezing anyway, so the question of freezing superchilled fish should not arise.

The amount of drip loss, loss of fluid from the thawed fish, will be greater than from fish properly frozen and stored at $-30\,^{\circ}C$. Also, the weight loss in wet fillets from superchilled fish will be higher than in fillets from ordinary iced fish. According to observations on cod and haddock, the weight loss in wet fillets from gutted whole fish superchilled to about $-2\,^{\circ}C$ over a period of 15 to 20 days will be about 3 to 6 per cent where the fillets are held in ordinary ice for approximately 15 h after filleting. The corresponding loss in fillets from ordinary iced fish is 1 to 2 per cent.

With proper stowage conditions and avoidance of stowage in ice at depths greater than 40 cm, or less for some species, there will be little or no weight loss in the whole fish during storage whether or not superchilling is employed.

The fishroom can be divided into two or more separate sections with individual temperature control. Each section or compartment is filled in turn and in this way it is possible to apply less refrigeration to the later part of the catch if desired. If only part of the catch is to be superchilled it may not be necessary to install the special refrigeration system for superchilling in all sections, depending on the periods of storage.

The view that storage life is extended if the fishroom is completely filled with fish and ice and then sealed tight is held by operators, both for ordinary iced storage and superchilling. There is an increase in concentration of carbon dioxide gas under these conditions, which may have some effect but will not be a major factor. The fishroom should be shut tight except during stowage, however, if only to prevent the introduction of heat from outside air.

Refrigeration plant

The plant required for superchilling is modest by comparison with plant for quick freezing because the cooling rate is slow and does not have to follow the ups and downs of catching rate so closely and because there is only partial freezing at a relatively high temperature. The relationship between temperature and power required for refrigeration is discussed in Chapter 7.

For a normal maximum catching rate of 24 t/day and a fish temperature of $-2^{\circ}C$, the required refrigeration capacity for superchilling will be not more than 45 kJ/s with an electrical power requirement of about 15 kW for the compressor. The corresponding values for quick freezing and cold storage will be roughly 150 kJ/s and 150 kW. Thus superchilling has been attractive in some cases where there has been a reluctance to make the bigger change and greater investment for freezing at sea in distant water vessels. Many existing vessels which employ melting ice are unsuitable for conversion to freezing at sea but may be suitable for conversion to superchilling. As far as the fishing vessel itself is concerned, the main object of superchilling and of freezing is the same; to enable the vessel to spend more time on the fishing grounds.

Superchilling with bulk stowage

Superchilling with bulk stowage of the fish in ice has been practised on a commercial scale, notably in warmer waters using a special installation for the fishroom. The fish is bulk stowed in pounds in the normal way except that the ratio of fish to ice need not be lower than two to one. The sides of the pounds are fitted with heat exchangers of stainless steel plate which form partitions about 1 m apart from side to side of the hold, except that they are broken in the middle by a corridor. The aluminium shelves, which are removable, are not directly refrigerated but assist in the cooling by conducting some heat to the coolers. With this in mind, in order to make the cooling as rapid and as even as possible throughout the fishroom, the shelves are not spaced more than 40 cm apart. All other surfaces of the hold, deckhead, tanktop, sides and bulkheads, are cooled by grids buried under the surface in a layer of cement over the insulation. The finished surface is entirely of stainless steel in 40 cm squares. The fishroom is easy to clean.

The plate heat exchangers and grids are supplied with chilled brine at a controlled temperature, just below the desired fish temperature during the cooling down period and then at the desired mean fish temperature. The brine chiller is a compact unit with refrigerant compressor and standby compressor, condenser, heat exchanger or evaporator, pumps and controls. The pumps are of modest capacity and contribute only a small amount of heat. For a temperature rise of 1 degC, the difference between the supply and return brine temperatures, at the required 45 kJ/s capacity in the example above, the pumping rate will be about 10 dm^3/s.

The ice provides a suitable medium in which to store the fish and, in the normal way, supplies rapid cooling of the catch down to 0°C as soon as the fish is stowed. Further cooling into the superchilling range and heat gains to the fishroom are taken up by the brine cooling system.

The main disadvantage of the bulk stowage in ice is the difficulty of discharge of the catch on landing. An increase in brine temperature for a period of hours, or days, before landing can help to loosen the congealed ice and make discharge less difficult. Even with bulk storage in melting ice in the ordinary way, however, discharge of the catch is usually a laborious procedure with few mechanical aids.

Boxing

The system described above has been adapted for boxes in order to gain the advantages of ease of discharge and handling. Collapsible boxes of aluminium are used in place of the shelves but otherwise the method on board is the same. The sides and ends of the box are hinged at the bottom and open outward so that there is no difficulty in removing and, if necessary, immediately breaking up the contents of the box. Boxing also gives an advantage in cases where the fish is to be held for a short time or transported after landing. Unless inspection is required, the contents need not be disturbed until the fish are required for processing and distribution.

Another system of superchilling, with which there has been comparatively little experience, is the cooling of boxes of iced fish in recirculated chilled air. With forced circulation of air and a uniform space of 1 cm or more between the boxes, the amount of cooling can be closely controlled. This method also can be used to superchill only part of the catch, of course, and the cost of the installation will be lower than the cost of the system for bulking.

Many of the points made in the discussion of forced convection cooling apply here, with the major exception that the cooling load will be higher with superchilling since the fish are to be partially frozen. The air temperature rise will have to be restricted in order to have reasonably uniform conditions. Again in the example given above, the pump will be replaced by the fan which should have a capacity of about 33 m^3/s for a given temperature rise of 1 degC. The fan power will be considerable, roughly 15 kW depending on the resistance of the stack of boxes, air cooler, ducting, etc., adding 30 per cent or more to the refrigeration load. Particular attention will have to be paid to the questions of dehydration and defrost. The air cooler will have to be defrosted regularly in order to maintain good performance. The dividing of the cooling load between several units will lend itself to an effective hot gas system.

The fisherman should not be expected to carry out stowage in a blast of chilled air. Even when only part of the catch is superchilled, therefore, the space for superchilling may have to be divided into two or more sections, filling each section in turn and shutting off the supply of chilled air where stowage is taking place.

Chapter

Refrigerated Sea Water 6

RSW/CSW

IMMERSION of the fish in refrigerated or chilled sea water has been used as an alternative to ice for the chilled storage of a wide range of species including salmon, halibut, tuna, herring, mackerel, blue whiting and shellfish. There are two methods of cooling; refrigerated sea water, RSW, in which mechanical refrigeration with a heat exchanger is used, and chilled sea water, CSW, in which ice is added to the water. The recommended ratio of fish to sea water is 2:1 to 3:1. The water is circulated through the fish, normally with no replacement of sea water, during the period of storage in order to cool the fish and prevent stratification. The recommended storage temperature is -1 to 0°C, just above the freezing point of the fish. Sea water has a salt content of about $3\frac{1}{2}$ per cent and a freezing point of -2°C, but there may be some dilution near the mouths of rivers.

The most important advantage of RSW/CSW over icing is the ease of handling and stowage on board, saving a great deal of labour. Indeed, in some fisheries where the fish are captured in large numbers over a short period, notably salmon and herring, it has not been possible to ice the catch properly.

Brailing is a widely employed method of unloading the fish out of the tanks on landing. Pumping systems have been developed for unloading fish of up to 10 kg at a rate of more than 25 t/h. In some cases the vessel is equipped with a pump for pumping the fish out of the sea into the holding tanks and this has been used for unloading. RSW/CSW storage can be extended to the fish on shore, in the same water if desired and without any significant increase in temperature.

As with icing, the addition of anti-bacterial substances to the water has not been found worthwhile. Such measures do not have much effect until spoilage is fairly well advanced. There has been some success with the addition of carbon dioxide gas, bubbled through the water, to shellfish and other fish but it has not been widely applied.The use of carbon dioxide gas will entail special safety precautions.

Proposals have been made for the use of containers of 1 to 2 t capacity for the storage and transport of fish chilled in CSW. Agitation can be provided by bubbling gas, nitrogen or carbon dioxide, through the contents of the container. The method is attractive because good storage conditions can be maintained during the voyage and after landing but it has not found favour because fishroom capacity is reduced and special arrangements are required to handle the containers and carry them on board.

Salt penetration

The penetration of sodium chloride salt into the fish and sometimes the bleached appearance of the skin or the fact that scales have been removed from the fish have discouraged the use of RSW/CSW in some fisheries but, except in extreme cases where the period of storage has been rather long, these do not affect quality, especially eating quality, to any great extent. The amount of increase in salt will depend on several factors; the size and species of the fish, whether or not they have been gutted, the ratio of fish to water and the length of period in storage. Gutted fish or fish which have been cut will tend to absorb salt at the cut surfaces. With the recommended ratio of fish to sea water of 2:1 and where there is no introduction of sea water after the initial stowage, the amount of salt taken up by the fish cannot exceed about 1 per cent by weight.

Increased concentration of sodium chloride to a total of 1 per cent or more, as opposed to the initial values of less than ½ per cent, can be reached in about 7 days in many species including cod, whiting and herring. The rate is lower in salmon, reaching 1 per cent in about 14 days, and much lower in halibut and tuna. A concentration of 1 per cent will be objectionable in many forms of fish products. If the fish are to be frozen and cold stored after a period of chilled storage, the permissible increase in salt content may be relatively low, since the salt can accelerate some of the changes which take place in cold store and hence decrease storage life. With gutted cod or hake for example, the limit may have to be set at 12 h. Fish that have been held for a substantial period in RSW/CSW also may be unsuitable for the production of fish meal, bearing in mind that the salt concentration will be increased about five times by drying. In the worst case, white fish to be dried without any mechanical separation of water, the normal limit for fish meal is ½ per cent in the wet fish. The problem may not be so serious where some of the water is extracted mechanically in a press or centrifuge after the raw material has been cooked. Storage in RSW/CSW for moderate periods is employed, however, on vessels supplying fish meal factories.

In most instances, however, the increase in salt content is of little significance. There are important cases where a substantial increase can be tolerated, especially where salt is added during processing after landing. Outstanding examples are in the canning of salmon, tuna and shrimp, and in the production of smoked products such as kippers from herring where the fish are dipped in brine before smoking. On the other hand, in some cases it will be feasible to substitute fresh water, wholly or partly, for sea water where clean sea water is not available or where the product demands it.

Spoilage

The advantages of adequate chilling are considerable, as emphasized in Chapter 1. They are illustrated to some extent in Table 6.1 which gives results for the machine splitting of herring. The split fish were assessed for their suitability for the production of kippers. They had been frozen whole, cold stored for a few weeks and then thawed before splitting.

Table 6.1 'Splitability' of herring

Treatment before freezing	*Time before freezing, h*	*First-class fish, per cent*
none	0	92
RSW (or iced)	32	75
unchilled	8	33

One would expect an appreciable increase in shelf life due to the slightly lower temperature that can be used with RSW/CSW but possibly because of a difference in the amount of aeration at the surfaces of the fish, it appears that the pattern of bacterial spoilage is not the same as it is in ice. In practice, where good procedures are followed, the rate of spoilage is much the same, however, sometimes a little lower in RSW/CSW storage.

There is some leaching of soluble components, including protein, from the flesh. These losses are not of serious proportions, however, and are generally only slightly higher than in iced fish. The development of rancidity during storage in RSW/CSW is not as great as in melting ice because there normally is less oxygen available in RSW/CSW. The tanks are usually filled completely in order to avoid instability of the vessel, hence the degree of aeration of the water is not very high. As in iced fish, severe breakdown of the belly wall will occur in ungutted fish after a time, depending on the amount of enzymic activity. Discoloration of the flesh in the region of the gut also can occur. Fish that have been feeding heavily will have a much shorter storage life. Thus the storage life of herring usually will be 6 days but only about 1 day when the fish have been feeding heavily and the storage life of salmon may be reduced from 14 days to 3 days.

With a few exceptions, fish stored in RSW/CSW tend to gain weight rather more than fish iced in boxes or in shallow layers. In most species the gain in weight will be in the region of 2 to 5 per cent. Whiting and cod and perhaps some other species will have a larger gain, however, more than 10 per cent over a period of 7 days. Shrimp can lose more than 10 per cent over a period of 12 days, but the rate in melting ice is twice as great.

THE RSW/CSW SYSTEM

Storage tanks

The tank should be watertight, easily cleaned and should not give rise to contamination of the fish. It should be designed with stability of the vessel in mind. Tanks of aluminium, fibreglass reinforced plastic and steel coated with protective and anti-corrosive substances have been used. Tanks constructed of marine plywood also have been used, notably in the Pacific salmon fishery in wooden vessels. The plywood is installed in two layers with staggered joints and fibreglass reinforced plastic or rubber based paint is applied to the inside. The space between the tank and the wooden hull usually is not insulated in order to allow any water which may leak out of the tank to drain away and to give a ventilated space so as to discourage the development of wood rot. The amount of heat gained through the uninsulated wooden hull is small in comparison with the total refrigeration requirement. Insulation is employed in steel vessels, of course, otherwise heat gains would be too high. Ideally the tank should be insulated from the metal frames of the vessel. With the tanks installed directly against the frames, but with insulation between the frames, the overall thermal resistance will be fairly low, typically 0·3 m^2 degC/W. This may be sufficient, however, especially for an RSW system. An insulation thickness of 5 cm over the frames will give a thermal resistance of about 1·4 m^2 degC/W. The bulkheads dividing the tanks are not normally insulated. Where insulation is installed, it is essential that the tank be perfectly watertight, otherwise leaks might give rise to a serious source of bacterial contamination.

The tank volume is divided into several parts in most cases, normally not less than three sections, each with a volume of up to 100 m^3. The tank should be filled completely with fish and water in order to prevent physical damage to the fish due to the motion of the vessel and to reduce aeration of the water. It is standard practice to reduce these difficulties and give stability to the vessel by constructing the tank with a narrow section or neck at the top. Partial filling of the tank must be avoided where it will create instability.

Circulation of the water for cooling is effective even with a fish to water ratio of 3:1. The water velocity over the fish should not be high but the circulation rate should be sufficient to ensure even temperature distribution throughout. Various methods of circulation of the water in the storage tank have been employed. The arrangement must take into account the tendency for the fish to block the area where the water is drawn out of the tank. Large suction screens, typically of expanded metal with openings of about 2 cm and an area of 0·25 m^2 per m^3 of tank, have been used to enable good circulation by means of a pump located outside the tank. In some systems for herring, agitators have been installed in the tanks, performing the task of fluidizing the mass of fish, rather than rely on the circulating pump for this purpose.

Circulation from bottom to top of the tank is preferred in most respects but top to bottom circulation has been used with pump systems because it permits circulation in a partly filled tank. According to one method which has given good results, however, a large suction screen is installed at one vertical bulkhead and the water is supplied to the tank through a distributor at the bottom. When the tank is only partly loaded the fish block the screen and the water is forced to flow through the mass of fish and over the top, through the open part of the screen to the pump suction.

The usual procedure on charging the tank is to admit fish to the tank after it has been partially filled with the appropriate amount of water and refrigerated. Spare fish tanks can be used to provide some reserve of chilled water before they are filled with fish but often it will be desirable to limit the total amount of water in order to keep salt penetration low.

Pumps and piping

Ordinary centrifugal pumps, installed below the level of the tanks where possible, are employed to recirculate the water through the tanks. The pump normally is arranged to discharge through the water chiller. Separate pumps may be installed for each tank or one pump may serve a number of tanks in parallel. The pump capacity should be at least 25 dm^3/s for each 100 t of fish to be stored.

Piping can be a source of bacterial infection if it is not properly designed and maintained in clean condition. Galvanized iron piping is commonly used. Plastic piping is most suitable except where it may suffer physical damage. It is advisable to arrange the piping in a closed circuit with a connection for the introduction of cleaning agents, so that the piping can be disinfected separately, and with a connection for flushing overboard. Pipe fittings should be chosen carefully. Simple ball valves and butterfly valves are preferred because they are easy to operate, have low resistance to the flow of water and are inherently relatively clean.

Mechanical refrigeration

In a properly designed system the amount of refrigeration required depends mainly on the initial temperature of the fish and the amount to be loaded into the RSW tanks. The load imposed by the heat gained through tank insulation, etc., and by the pump is relatively small although significant. In all cases it is very important to ensure rapid cooling of the catch to a temperature below 5 °C. The storage temperature should be maintained at slightly below 0 °C, preferably −1 °C, but freezing of the fish should be avoided except where superchilling is to be employed.

Thus there are considerable variations in the required cooling capacity, depending on the fishery and other circumstances. Often the cooling load is high, because the RSW method is used primarily to handle large catches with a minimum of labour. As an example, a catch of 100 t at a temperature of 20 °C will impose about 8,000 MJ on the refrigeration system. To remove this amount of heat in, say, 6 h would require a large plant with a power requirement of more than 100 kW for the compressor(s).

Chilling in RSW can be somewhat faster than chilling in melting ice. In practice, however, cooling is not always faster because of limited capacity of the mechanical system.

The cooling of some sea water before fishing begins is usually helpful since it enables quicker cooling of the catch. This can be done by filling one or more tanks with water, taking care to ensure stability of the vessel. A typical arrangement for an RSW system is shown in Fig 6.1.

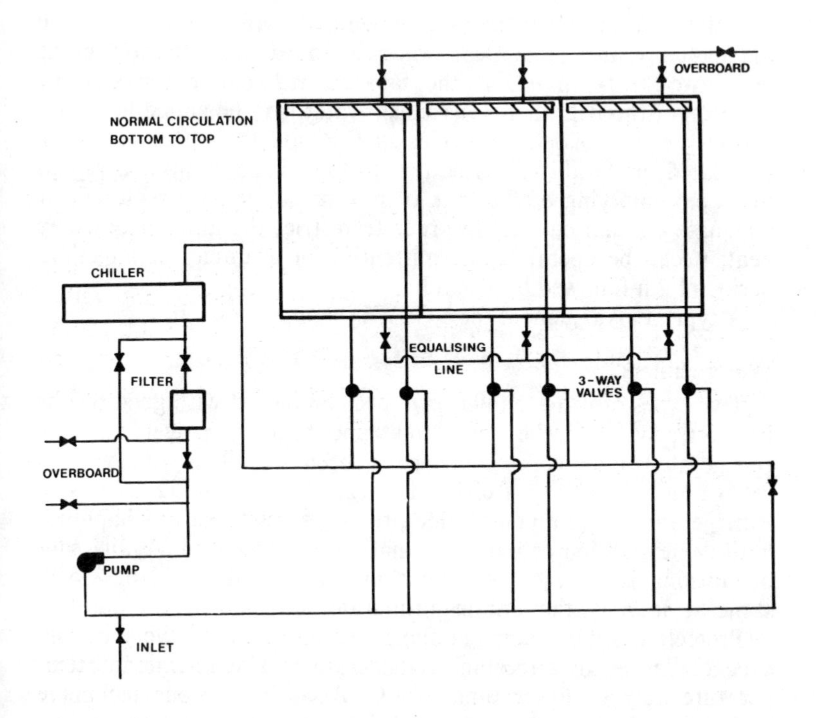

FIG 6.1 Piping arrangement for RSW system

Ice

The addition of ice directly to the storage tank can be used as an alternative to, or to supplement, the mechanical system. It is often the preferred method, especially for the shorter trips of five days or less, because it is simple and deals effectively with the high initial cooling load. In the example above the initial load would be met by the addition of 24 t of ice. Supplies of ice from shore usually are preferred to the installation of a flake ice machine on board. Freshwater ice will give a dilution of the salt which is an advantage where salt penetration might be a problem, although the freezing point of the water will be raised. The presence of ice in the storage tank does not introduce any difficulties with circulation; it is possible to pump mixtures of flake ice and water with the water content as low as 10 per cent.

According to one recommended procedure the tank is charged with ice before the beginning of the trip. Just before the fish are put into the tank any meltwater is pumped overboard, the ice is broken up manually and some clean water is added, leaving the tank at least two-thirds empty. If the tank is not completely filled by addition of the fish, then more clean water may be added to ensure stability of the vessel. After closing the tank the water should be circulated, preferably from bottom to top. If the pump is designed for rapid emptying of the tank it may be larger than required for adequate circulation and, in order to restrict the amount of pump heat, it can be operated intermittently, for example for an initial period of 2 h followed by 1 in 3 h.

A typical arrangement for a CSW system is shown in Fig 6.2.

Water chillers

Two types of water chiller have been employed with good results in RSW systems; the shell and tube heat exchanger and the refrigerated pipe coil. These have been installed with various refrigerants including Refrigerant 12, Refrigerant 22 and with Refrigerant 717 (ammonia) along the lines described in Chapter 7, with dry expansion or pump circulation of refrigerant. Again, ease of cleaning is a major consideration because there will be blood, slime, scales and other debris in the water.

Protection against damage due to ice formation in the shell and tube chiller is an important consideration. The refrigerant temperature may be lower than −5 °C. According to one technique which has been employed in the Pacific salmon fishery, the refrigerant is in the tubes (dry expansion) and the water is on the outside. The shell is of plastic in order to avoid rupture should there be accidental freezing and because of its resistance to corrosion. Copper tubes have been found satisfactory, with a life of more than 10 years, although it is recommended that the water velocity over the tube should not exceed 60 cm/s, otherwise the copper may suffer corrosion from impingement attack. Copper cannot be used with ammonia refrigerant, of course, because it is readily attacked by ammonia. Chillers of galvanized steel also have been satisfactory. Usually several small chillers are installed in parallel, giving maximum flexibility and reliability, connected to inlet and outlet headers by short lengths of rubber hose. A filter with provision for back-flushing the filter screen is placed in the pump discharge line to protect the chillers from debris in the sea water.

The cooler surface required with dry expansion of refrigerant is 36 dm^2/kW with a temperature difference of 5 degC between sea water and refrigerant and a water velocity of 30 cm/s (heat transfer coefficient 560 W/m^2 deg C).

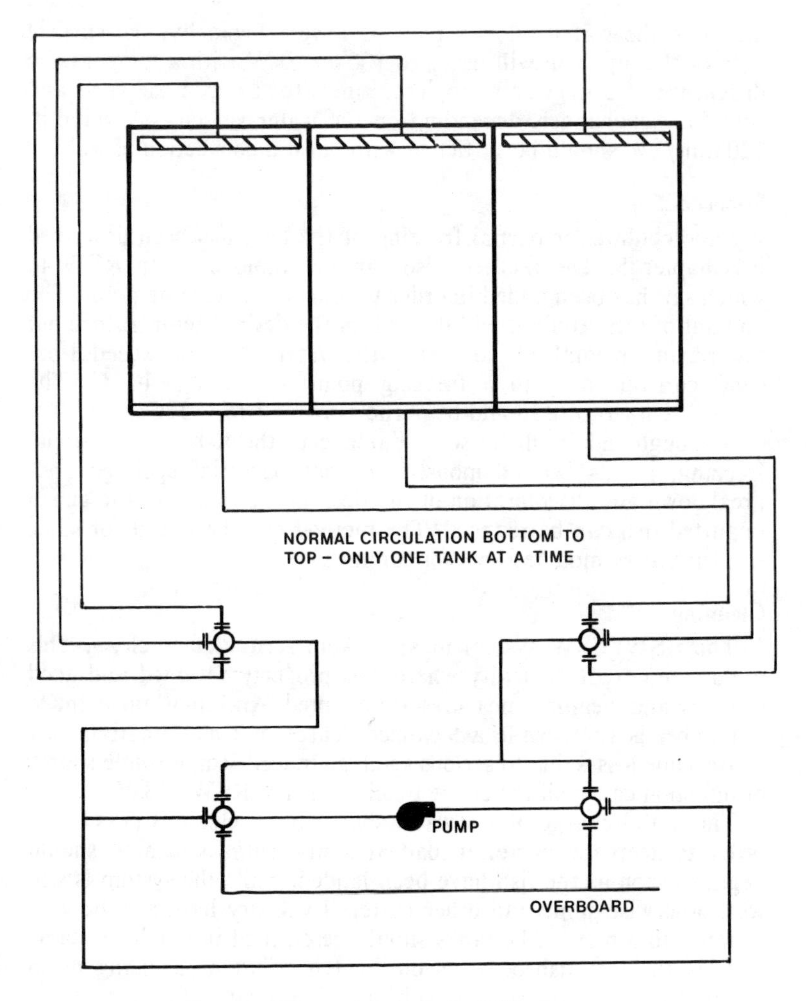

FIG 6.2 Piping arrangement for CSW system

The pipe coil can be made of galvanized steel tube with an outside diameter of 3 to 5 cm, installed in a separate tank through which the sea water is circulated, or it may be installed in a partitioned section of the fish tank. It is robust and there is little danger of damage due to freezing of water on the coil. This method consumes more space than the shell and tube heat exchanger but fundamentally it is similar, with the sea water outside the tube, although the water velocity may be lower. Typically, the surface area of the pipe coil will be 70 to 100 dm^2/kW with a temperature difference of 5 degC, two to three times the area in the shell and tube heat exchanger, depending on the water velocity. A value of 120 dm^2/kW should be sufficient with natural convection of water.

Superchilling

Superchilling, or partial freezing of the fish, has been described in Chapter 5. The method also can be employed with RSW to which salt has been added in order to reduce the freezing point. The amount of salt required will depend on the desired temperature but the maximum total salt content of the water need not exceed 8 per cent, corresponding to a freezing point of about −4·5 °C. The applied temperature should be in the range −1 to −3 °C.

Although there will be some damage to the fish due to partial freezing, it has been demonstrated that bacterial spoilage and-breakdown and discoloration of the flesh in the region of the gut in ungutted fish can be reduced. The method may be useful for some fish such as salmon destined for canning.

Cleaning

The RSW/CSW system must be kept scrupulously clean. This imposes no great difficulty where it is properly designed and good sanitary and cleaning measures are adopted. An important point to remember is that, while a localized source of contamination may cause some losses due to serious spoilage in iced fish, a single source of infection can spoil the entire batch of fish in RSW/CSW.

The initial charge of sea water should be as clean as possible in order to keep the bacterial load at a minimum. Cleaning should begin as soon as the fish have been landed, while the system is still wet, otherwise slime and other material will dry hard and be very difficult to remove. The tanks should be cleaned in much the same way as the wet fishroom, as outlined in Chapter 3, using clean water from a hose and a suitable detergent, brushing where necessary to remove any material which has adhered to surfaces. The piping system, including pumps and heat exchangers, should be flushed out thoroughly and then cleaned by circulating hot water or a cleaning solution. Sometimes a weak solution of disinfectant is left in the piping system until just before the tanks are to be used again, and then the entire system is thoroughly flushed with clean sea water before it is put into use.

Chapter

Refrigeration Plant 7

Plant for the Fishing Vessel

A DISCUSSION of the fundamentals involved in the design and operation of refrigeration plant is not appropriate here. There are a number of textbooks on this subject. Of course the larger plant, especially for freezing and cold storage, should be under the control of a skilled operator but in any event there should be an appreciation and knowledge of the application of refrigeration plant to the particular problem of the refrigeration of fish at sea.

As with much other apparatus on board, simplicity and reliability are of prime importance. The high inertia forces due to the motion of the ship, the vibrations encountered, the corrosive effects of sea water and the often robust nature of the work make conditions difficult for the designer. Delays due to breakdowns and malfunctioning of the plant can result in serious loss of fishing time and poor quality. The vessel may be far from a port where a fault can be rectified and the manpower and facilities for servicing at sea obviously must be limited as far as possible.

Added to this, the plant, including freezers where they are employed, must have a high performance so that it is compact. Space on board is valuable and at the same time high output is required, especially during periods of heavy fishing.

Vapour compression

Virtually all mechanical plant for the refrigeration of fish at sea is of the vapour compression type with sea water condensers, sometimes with compound compression and intercooling. The usual refrigerants are Refrigerant 717 (ammonia), Refrigerant 12, Refrigerant 22 and Refrigerant 502. Ammonia is relatively cheap and has considerable thermodynamic advantages. It is the most widely used, therefore, but in a number of countries there are severe restrictions placed on its use on board ship, mainly because of explosion risk and because of its toxicity. There are differences of opinion on the degree of danger. Few accidents have been reported. Leaks of ammonia to atmosphere are invariably quickly detected and repaired because the smell is so strong, even at very low concentrations. Leaks of Refrigerants 12, 22 or 502 are not so readily detected, on the other hand, because there is no smell but there are suitable leak detection instruments on the market. It should be borne in mind that, because the refrigerant vapour will displace the normal atmosphere, leaks in a confined space can be particularly dangerous. Also some doubts exist over the degree of risk to health from exposure to these refrigerants and special safety precautions are recommended by the suppliers. The prevention and detection of leaks are always important points and must be given particular emphasis on board ship from the points of view of both servicing and safety within the confines of the ship.

Refrigerants 22 and 502, although higher in price, are often preferred to Refrigerant 12 for freezer installations because the required compressor displacement volume is lower, some of the pipes can be smaller and the operating pressure on the suction side of the compressor is higher. The Refrigerant 22 system operates at above atmospheric pressure down to about $-40\,^{\circ}C$ whereas the pressure at $-40\,^{\circ}C$ in the Refrigerant 12 system is about 630 mb absolute. Thus there will be less trouble from ingress of air into the Refrigerant 22 system due to leaks in the suction side.

Compressors

The reciprocating compressor is used in most installations, sometimes with a rotary booster in the low stage in low temperature applications. It is the most suitable type at present but eventually may be superseded by alternative types. The screw compressor has been improved in recent years. Problems of noise and of recovery of oil in the discharge gas have been largely solved and space saving arrangements in which the screw is positioned vertically have been developed.

In the smaller fishing vessels, the compressor often is driven by an internal combustion engine which also may drive the condenser sea-water pump. In the larger vessels electric motors are used. In any case there should be the usual pressure switches to stop the compressor in case of failure of the condenser or other difficulties. The compressor lubrication system must function properly under conditions of rough weather.

Some form of capacity control is essential because of the wide variations in load normally experienced. With small units, starting and stopping of the compressor usually offers no serious complications. Various methods are employed in the larger installations, particularly for freezing at sea, including the installation of multiple compressors, cylinder unloaders, speed control and controlled bypass of hot gas from discharge to suction. Sometimes more than one method is used. A multiple installation is desirable for reduction of the starting load imposed on the ship's electrical supply system.

Normally there should be more than one machine, or set of machines, so that there is standby capacity in case of breakdown. Since a small capacity is required for the holding of frozen fish, as opposed to freezing, not more than one small unit need be installed for the cold store, with standby capacity provided by the main compressors for freezing.

Typical performance characteristics of reciprocating machines are shown in Fig 7.1.

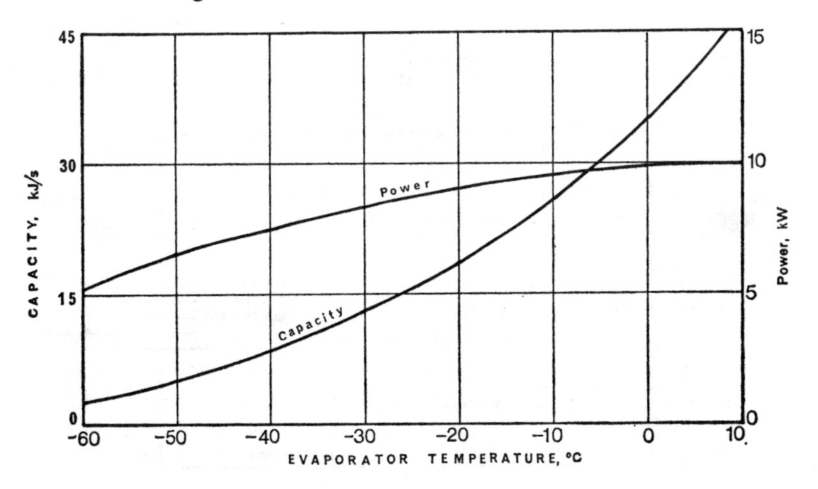

FIG 7.1 Typical performance characteristics of a reciprocating compressor

Condenser and receiver

The shell and tube heat exchanger with sea water as the condensing medium is the most common type of condenser. As in all properly designed systems, there should be ample condenser surface. The condenser cooling capacity should be more than double that required at the normal design condition, to allow for fouling and for high load during cooling down periods. In selecting a condenser, consideration should be given to the prevention of corrosion and facilities for cleaning and flushing out the condenser.

The capacity required of the condenser sea water pumps is not high. For a water temperature rise across the condenser of 2·5 degC at the normal running condition, 1 dm^3/s will be required for each 10 kJ/s to be rejected. A standby pump will be a worthwhile investment.

A separate liquid receiver should be installed below the condenser in all but the smallest of systems. It can serve to stabilize conditions in the refrigerant circuit and accommodate the charge of refrigerant when the rest of the system is pumped out for repairs or alterations. An upright receiver is preferred on board ship.

REFRIGERATION SYSTEMS

The basic mechanical refrigeration system with a 'dry expansion' evaporator, or cooler, is shown diagrammatically in Fig 7.2. Heat transfer conditions in the evaporator are much less than ideal due to

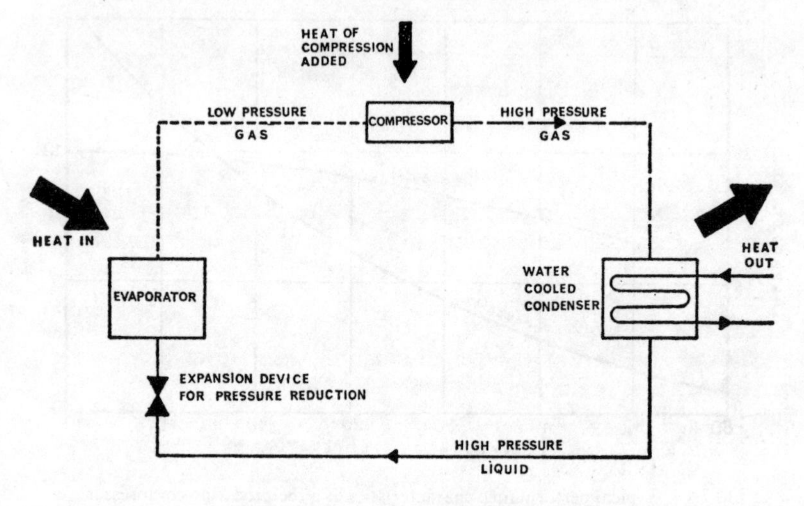

FIG 7.2 Mechanical refrigeration system

poor wetting conditions inside but sometimes this method is chosen on account of simplicity, especially in small systems and where there is only one evaporator. Where better heat transfer conditions are required and in instances where there otherwise would be numerous expansion devices on a number of evaporators in parallel, 'recirculation' or 'overfeed' systems in which the rate of flow of refrigerant is higher than the evaporation rate are employed. Circulation can be by gravity, not normally recommended, or by pump as shown in Fig 7.3.

The pump system usually is preferred and has been adopted widely for fish freezing installations because of its simplicity,

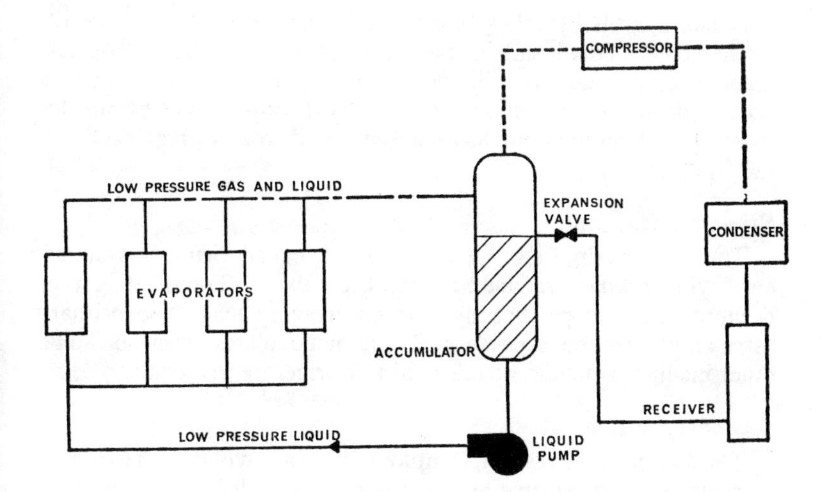

FIG 7.3 Pump circulation, primary refrigerant

reliability and good heat transfer performance. There is no need for a controlled expansion device at each cooler. Since the circulation rate is higher than the evaporation rate, there will be a mixture of gas and liquid returning to the separator, sometimes called the surge vessel or accumulator.

The heat transfer and pressure drop characteristics of the system should be analysed at the design stage in order to determine the pump load and suitable values of liquid flow, pipe sizes, etc. The geometry of the evaporator and the return lines to the separator is particularly important because with the two phases, liquid and gas, pressure drops tend to be high and temperatures in the evaporator uneven. Thus there is a balance to be struck between heat transfer and pressure drop. As an indication, however, the rate of liquid flow is commonly set at between two and six times the evaporation rate at full cooling load. Sometimes the rate is set higher than six, high enough to ensure good distribution between a number of evaporators, although the optimum value based on purely heat transfer and pressure drop considerations may be as low as two.

In the discussion above it has been assumed that evaporation will occur in the cooler. This need not be the case. By increasing the rate of liquid circulation to well above the normal range, more than 20 times the evaporation rate, little or no evaporation occurs in the cooler and the heat transfer characteristics are improved. The resistance to flow and power for pumping can be kept within reasonable limits by enlargement of flow areas and parallel circuits in the cooler. This is called 'flash cooling' because the appropriate amount of refrigerant will evaporate in the separator, under a sudden decrease in pressure. This method may be worthwhile for freezing installations but has not been used to any great extent in the fish industry.

Primary refrigerant

The systems in Figs 7.2 and 7.3 are called *primary* systems, employing *primary refrigerant*, because only one liquid refrigerant is employed to cope directly with the cooling load. The primary refrigerant may be used to cool air, metal plates, brine or some other medium which in turn is used to refrigerate the fish.

Secondary refrigerant

The *secondary* system, employing a non-volatile *secondary refrigerant* as shown in Fig 7.4, has been installed in a number of vessels, mainly for reasons of reliability. In the event of an undetected leak in a primary system, large quantities of costly refrigerant can be lost. With a suitable inexpensive secondary refrigerant circulated at modest pressure, the charge of primary refrigerant is greatly reduced and confined to a smaller region with a reduced amount of pipe and fittings. The secondary system normally will be more costly to install and it has a disadvantage in that there must be a temperature difference, normally about 5 degC, at the heat exchanger or evaporator. Thus larger compressors and greater power input to the compressors are required because of the lower suction temperature.

The choice of secondary refrigerant is a matter of compromise. There is no entirely suitable fluid. The viscosity of glycols at low temperature is too high for good heat transfer and the fire risk with alcohol can be high. Trichloroethylene has been used as a secondary refrigerant in plants for freezing at sea and is preferred over methylene chloride which also is used as a secondary refrigerant. Trichloroethylene has greater toxicity but a lower vapour pressure so that, in the event of a leak out of the system, it will not vaporize so rapidly. The concentration at which one begins to smell trichloroethylene in the atmosphere is about 50 parts per million and, as a general rule, the concentration should not be allowed to rise to this level, although it is reckoned to be somewhat below the danger level. Trichloroethylene is not easy to contain, as it dissolves rubber and jointing compounds. Mechanical jointing is used.

Calcium chloride brine is also used but is impracticable for

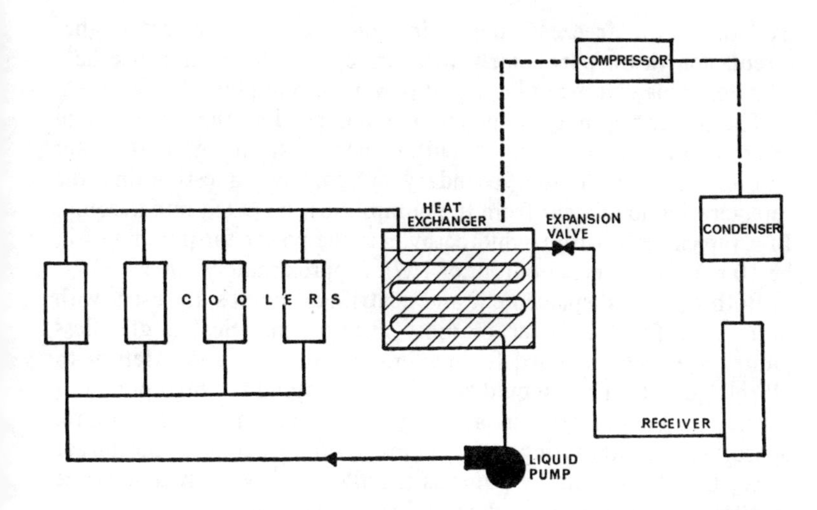

FIG 7.4 Pump circulation, secondary refrigerant

temperatures much below $-35\,^{\circ}C$ because of high viscosity which implies poor heat transfer. At $-40\,^{\circ}C$ the required area of cooling surface in the evaporator is relatively great, more than twice the area required with trichloroethylene. It has the advantages of safety and cheapness, however, and applied temperatures in the region of $-35\,^{\circ}C$ are feasible in some freezing installations.

Pumps

The required pump capacity is directly related to the cooling load and the need to maintain velocities high enough for efficient heat transfer. Excessive capacity will mean that power for pumping and the sizes of the pump and the pipes will be greater than necessary. The pumping rate to freezers in a primary system is normally not more than 5 dm^3/s for each 1 t/h of freezer output. This corresponds to about 15 times the mean evaporation rate for Refrigerant 22 and 90 times for ammonia. In simple installations the rate need be only about 5 times the evaporation rate but this may be too low for good distribution of refrigerant where there are a number of freezers in parallel.

The rate required with secondary refrigerant is higher because there is only sensible heating of the fluid. Based on a mean temperature rise of 2·5 degC, the rate for trichloroethylene should be 20 dm^3/s for each 1 t/h and the rate for calcium chloride brine should be about 12 dm^3/s. It might be desirable in some heat exchangers or freezers, depending on the type, to use higher circulation rates, particularly with brine, in order to improve heat transfer. This will mean increased power for pumping.

The power required by the pump will depend on the capacity and the resistance to flow in the circuit. Generally speaking, more power will be required in the secondary system. When estimating the refrigeration load, heat from the pumps must be taken into account. In a typical case with trichloroethylene, the power for pumping will be 10 kW for a freezer output of 1 t/h (approximately 100 kJ/s).

Both positive displacement and centrifugal pumps are used, with preference for the latter in most installations. Sealed glandless pumps are recommended for maximum reliability and safety with all refrigerants. They should be able to withstand the high pressures sometimes encountered in a refrigeration system. Sealed pumps, operating on alternating current, are used in many of the larger ships. Usually a standby pump is installed and possibly a separate smaller pump for the cold store in freezing systems.

A particularly important point with primary refrigerants is that the static head of refrigerant at the pump suction must be sufficient to prevent the vaporization of the refrigerant at the pump. Although head room tends to be limited on board ship, this seldom poses a problem where care has been exercised in the choice of pump and in the arrangement of separator, pump and piping between the two. The entry to the pump should offer as little resistance to flow as possible and no vapour should be entrained in the liquid to the pump. Special entry devices can be installed in the separator to eliminate vortex formation and bubble entrainment in the separator. The optimum velocity of the refrigerant in the suction pipe to the pump is 1 m/s.

Hot defrost

Periodic defrosting of heat exchangers such as air coolers and freezers is necessary in order to maintain good heat transfer. The accumulation of frost on freezers can interfere with loading and unloading of the fish. There are various sources of heat available for this purpose; water direct from the sea or heated water piped to the frosted surface, electric heaters attached to the heat exchanger itself, hot gas at high pressure from the compressor and, in a secondary system, hot secondary fluid heated from any convenient source including hot gas, steam and electricity. Regardless of the method, it should be noted that the hot defrost normally will impose a load on the compressor plant, sometimes as much as 20 per cent of the total load, depending on the heat in the heat exchanger, associated piping, etc., after defrost.

The use of hot gas is most attractive because, apart from some heat from compression, the source of heat is the refrigeration load itself and this heat otherwise would be rejected at the condenser. It is most effective in a multiple system with a reasonably large number of heat exchangers or evaporators so that a substantial amount of hot gas is available for the rapid defrosting of one unit at a time. In some cases it is advisable to limit the defrost temperature by the use of desuperheated vapour from the top of the condenser rather than superheated vapour direct from the compressor.

Piping and accessories

The arrangement and sizes of pipes in the refrigeration plant must be in accordance with good standard practice. With pump circulation of refrigerant to a number of freezers or heat exchangers in parallel, the piping arrangement should ensure that there is even distribution and that the resistance to flow of refrigerant, hence power required by the pump, is not unduly high. In that part of the primary system where there is a mixture of liquid and gas from the evaporator, usually at a mass flow rate of at least five times the evaporation rate, the resistance to flow tends to be relatively high, so pipe sizes should be chosen with care. Valves, preferably with the handles normally removed, and other devices can be used to obtain fine adjustment of distribution. The problem would appear to be more difficult in the primary system because resistance to flow will be higher in the coolers with higher evaporation rate, creating bad distribution and impairing heat transfer. In practice, however, there has been no difficulty in systems with carefully designed piping. Pipe insulation should be used, and the insulation for cold pipes should be sealed against the entry of water vapour.

It is good policy to choose fittings and accessories such as valves, oil separators, flexible pipe connections where required, expansion devices, dehydrators, liquid line heat exchangers and strainers mainly on grounds of reliability. Sight glasses should be installed at appropriate points in order to have a visual indication of the refrigerant in the system.

Controls

All the usual refrigeration controls have been used at sea; compressor capacity controls, compressor cut-out controls for low suction pressure, high discharge pressure, low oil pressure and motor overload, high pressure relief valves and bursting discs, thermostatic and hand expansion valves, solenoid valves for control of liquid and gas flow, automatic back pressure valves which may be used to maintain different evaporation temperatures in more than one evaporator served by one compressor, thermostats of various kinds and three-way mixing valves in secondary systems.

Automatic controls are essential for safety, accuracy and ease of operation but often too much reliance is placed on them. As a precaution against breakdowns, hand valves and other alternative manual controls should be included in the system. Ball valves with seats of plastic (polytetrafluorethylene) have proved to be reliable for both cold and hot fluids in pump circulation systems. Because of their ease of operation and low resistance to fluid flow they are suitable in pump suction lines and for changeover from cold to hot defrost fluid and vice versa at the freezer.

The liquid primary refrigerant in a separator or in the evaporator of a secondary system must be maintained at the required level. Various devices are available for this crucial function but perhaps the two most important are the float control and the thermostatic expansion valve.

A float in the separator or evaporator can be used to operate a needle valve through a simple lever mechanism, admitting liquid refrigerant as required to maintain the level. An alternative method is to operate a switch from the float, controlling an electric valve in the liquid line.

The thermostatic expansion valve with heated control bulb is a good method. The bulb is situated in the separator or evaporator at the required liquid level and the valve in the liquid line. A small amount of heat is applied to the bulb, usually by an electric element but sometimes by a little warm refrigerant vapour from the high pressure side of the system. Thus when the bulb is not submerged its temperature will be high and the valve will open but when it is submerged in liquid the temperature will be low and the valve will close.

The level control also can be placed in a liquid receiver in the high pressure side of the system, maintaining a given level in the receiver by metering liquid to the separator or evaporator.

Chapter

Freezers 8

TEMPERATURE AND FREEZING TIME

IN order to freeze fish, it must be placed in contact with something cold. Generally speaking, it will be necessary to bring the fish down to $-30°C$ in the freezer and then store it at that temperature. To achieve this, the temperature applied in the freezer should be below $-30°C$. It is good practice to employ a temperature of $-40°C$.

Of course there will be a temperature gradient in the fish while it is freezing. The outer layers will freeze first and will tend to approach the applied temperature soon after freezing begins, while the inside remains unfrozen for a time. Eventually, as more and more heat is extracted and the layer of frozen material through which the heat is extracted becomes thicker and thicker, water will be frozen throughout the material. At this point the temperature of the last part to freeze will be higher than $-30°C$, say $-20°C$, and the temperature of the outer surface of the fish normally will be lower than $-30°C$. Thus the mean temperature of the fish will reach the desired $-30°C$ before the last part does, usually before the last part reaches $-20°C$. It is common practice, therefore, to consider freezing complete when the temperature at the warmest part, which can be measured, has reached a value of $-20°C$. It should be emphasized, however, that the mean temperature of the frozen fish before storage should not be higher than the cold store temperature. Also from this point of view delays after freezing, during which the fish can warm up before stowage, can have serious effects on quality and should be strictly avoided.

Temperature measurement

The temperature of the last part to freeze, hence the freezing time, can be measured with the aid of a suitable thermometer, commonly the thermocouple or the resistance thermometer probe used with the appropriate indicating or recording instrument, inserted into the fish before it is frozen. In the thermocouple system, the thermoelectric characteristics of dissimilar metals are employed to measure temperature. The temperature sensitive junction is made by joining together the ends of metal wires. It is sensitive, cheap, reliable and robust and has the advantage of quick response. Since the thermocouple end is not easily removed from the frozen material, usually the electrically insulated leads to the indicating or recording instrument are disconnected or cut, possibly to be connected again for measurement of fish temperature during cold storage and refrigerated transport and recovered after the fish is thawed. Thermocouples of copper-constantan wire are suitable for the range of temperatures and conditions encountered in the refrigeration of fish.

The thermocouple or probe should be inserted carefully into the fish or pack so that the sensing end will be in the last part to freeze. Errors from conduction of heat along the wires or probe to the end should be minimized by installing the thermometer with an appreciable length, usually 10 cm is sufficient, of the insulated wires or the probe at substantially the same temperature as the end. In a whole roundfish, for example, the thermocouple would be installed securely in the thickest part of the fish with the leads running for some distance along the backbone toward the tail.

Often it is necessary to assess the performance of freezing plant by measurement of freezing time. The freezing time must be known in order to set the rate of output of a freezing plant as high as possible and at the same time avoid the danger of insufficient cooling.

A typical record of fish temperature, giving the freezing time, is shown in Fig 8.1. The shape of the temperature curve at the last part to freeze should be noted. There is a pronounced pause, called the *thermal arrest*, at the point where most of the water is frozen, in the region of -1 to $-2\,^{\circ}$C depending on the product, followed by a steep slope toward the end of the freezing period. Since very often the freezing will take place unsymmetrically, the geometrical centre of the fish or pack will not necessarily be the last part to freeze. Thus the placing of the thermocouple in order to measure freezing time is to some extent a matter of judgement. Whether it has been well placed can be seen, however, by the shape of the freezing curve. In any case a number of records should be made in order to ensure the correct determination of freezing time.

The temperature of a block of frozen fish also can be measured by drilling a hole and inserting a thermometer. Unless good procedure is followed there will be gross errors. The thermometer must be accurate and quick to respond. A thermocouple or resistance thermometer probe is suitable; a glass thermometer is not. The hole should be neat, only slightly larger than the probe,

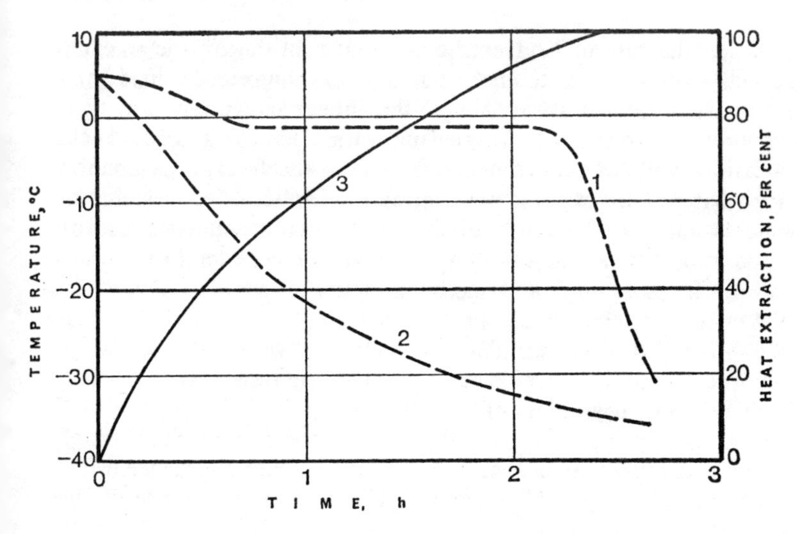

FIG 8.1 Temperature and freezing rate
1 centre temperature
2 temperature near surface
3 heat extraction

and at least 10 cm deep. The whole operation must be carried out quickly if the correct result is to be obtained under changing conditions, for example on removal from the freezer or cold store. With good technique, the minimum reading obtained soon after the probe is inserted will be less than 1 degC above the true temperature.

Rate of freeezing

The freezing time is dependent on a number of factors including the heat content, thermal conductance and shape of the material. Smoked or dried fish will have lower values of heat content and thermal conductance than wet fish. A single round fish will have a shorter freezing time than a large slab or solid fish of the same thickness.

An example of the rate of heat extraction is given in Fig 8.1. It is only a typical result. The exact shape of the curve will depend on the conditions. For practical purposes the rate of heat removal is considered to be the freezing rate but in fact both sensible and latent heat are removed as shown in Fig 2.1. The amount of sensible heat to be removed depends to some extent on the initial temperature.

When the rate of heat extraction is uneven during the freezing period, there tends to be variation in compressor load and evaporation temperature along the lines shown in Fig 7.1, evaporation temperature decreasing with decreasing load. These variations will be more pronounced where the freezing is done in large batches as opposed to continuous freezers and systems in which a number of batch freezers are operated in sequence. Steady evaporation temperatures often are desirable in order to maintain uniform freezing times and maximum output.

A useful, if somewhat imprecise, concept of the freezing process in fish is given if we consider that there are two resistances to the removal of heat; the surface resistance and the resistance of the fish itself. At the beginning of the freezing period the outside will freeze. As freezing progresses the layer of frozen material through which heat must pass becomes thicker and thicker and hence offers more and more thermal resistance. The thermal resistance of the frozen fish is only about 40 per cent of the resistance of unfrozen fish as given in Table 1.1, depending on the exact nature of the material, but nevertheless it is substantial. There also may be an appreciable resistance to the removal of heat from the surface of the fish, indicated by some values of surface resistance in Table 1.1, for example the value of 0·03 m^2 degC/W for air at 7 m/s. If freezing is to be as rapid as possible, the thickness of the item to be frozen and the surface resistance must be kept as low as possible.

The rate of freezing is dependent directly on the temperature difference across these two resistances. Since most of the water in the fish freezes in the zone just below 0 °C, the rate of freezing with an applied temperature, say, of −80 °C will be approximately twice the rate with an applied temperature of −40 °C, all other things being equal.

Summarizing these points, the heat extraction rate is directly proportional to the temperature difference and inversely proportional to the sum of the resistances, thus

$$H \propto \frac{D}{R_s + R_f}$$

where H = heat extraction rate at any instant
D = temperature difference
R_s = thermal resistance at the surface
R_f = thermal resistance of the frozen layer

R_s usually remains fairly constant during the freezing period. If its value is high compared with R_f, the heat extraction rate, or freezing rate, will be low and fairly constant. R_f becomes higher as the frozen layer becomes thicker. Thus, when R_s is relatively low as it often is in freezers for freezing at sea, the rate of heat removal is more or less along the lines shown in Fig 8.1, higher during the early part of the freezing period and slowing down toward the end. When R_s is negligible the freezing time varies as the thickness squared.

Based on the properties of lean white fish, including those given in Chapter 1, freezing times for two ideal shapes, the slab with heat extraction from two sides only and the cylinder with heat extraction from the curved surface only, are given in Fig 8.2. The initial fish temperature has been taken as 0°C and the final temperature −30°C (−20°C at the centre). The figure is based on the expression

$$KDT = 420\,LR_s + 0{\cdot}60\,L^2$$

where
- T = freezing time, h
- L = thickness, cm
- R_s = surface resistance (recriprocal of the heat transfer coefficient), m² deg C/W
- D = temperature difference, degC
- K = constant: 1 for slab, 2 for cylinder

Because Fig 8.2 is for ideal cases it cannot be applied with precision to real cases. The best approach is to measure freezing time. Fig 8.2 can serve as a rough guide in some cases, however, where the solid rectangular block of fish resembles the slab, the single roundfish resembles the cylinder and the flatfish lies between the two. For a case in which the freezing time has been measured, R_s can be estimated from Fig 8.2.

As an example typical of the horizontal plate freezer, the surface resistance, R_s, with a block of fillets will be 0·017 m² degC/W. For a thickness of 5 cm and $K=1$,

$DT = 51$ from Fig 8.2.

With a plate temperature of −40°C,

$T = 51/40$

$T = 1{\cdot}3$ h

As an example typical of the air blast freezer with an air velocity of 7 m/s, the surface resistance will be 0·030 m² deg C/W. For a roundfish with a thickness of 12 cm and $K=2$,

$DT = 119$ from Fig 8.2.

With an air temperature of −35°C,

$T = 119/35$

$T = 3{\cdot}4$ h

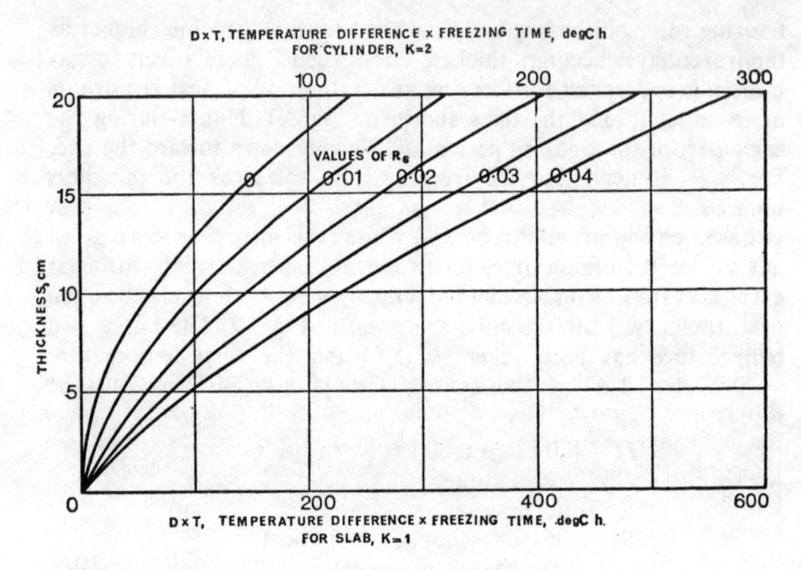

FIG 8.2 Freezing times

The freezing time will be increased by high initial temperature. The freezing time at an initial temperature of 20°C will be roughly 10 per cent longer than the time at 0°C.

Low temperature

Since the freezing rate depends directly on the temperature difference, the size of the freezer for a given output and hence the overall size of the refrigeration plant can be reduced by a reduction in the applied temperature. This is attractive because the saving of space on board ship may reduce the first cost and some of the running costs of the vessel. Temperatures much below −40°C have not been used, however, because their production is generally more difficult and less economic. The risk of brittle fracture in ordinary mild steel becomes an increasing problem at temperatures below −6°C. The reduction in suction pressure brings a reduction in compressor capacity, hence increased power input per unit of refrigeration as illustrated in Fig 7.1. Sealing against leaks in the suction side becomes more difficult.

The cascade refrigeration system is probably the best choice for the production of temperatures below −60°C but it is unlikely to find wide application on fishing vessels because of its relative complexity. In the cascade system there are two or more separate circuits at different temperature levels with the evaporator in the circuit at higher temperature acting as the condenser in the circuit at lower temperature. In a typical system, Refrigerant 22 at a temperature of −30 °C would be employed as the condensing medium for Refrigerant 13 operating at an evaporator temperature of −80 °C or below. Suction pressures would be positive in both circuits but there would be no great advantage on power input.

Low temperatures are not widely employed in the fish industry except for tuna as mentioned in Chapter 1.

Quick freezing

The term *quick freezing* is commonly used. It is defined in different ways in the various countries. According to one definition for lean wet fish, the fish is quick frozen if it is reduced in temperature from 0 to −5 °C in not more than 2 h and the fish is retained in the freezer until the temperature of the warmest part is reduced to −20 °C or lower. This usually implies a total freezing time to a mean temperature of −30 °C of not more than 4 h with an applied temperature of −35 °C or lower.

It is not possible to quick freeze very large fish according to this definition. With the usual applied temperature of about −40 °C and very low surface resistance, R_s, the overall freezing time of an item 15 cm thick will be less than 4 h, depending to some extent on the shape and the initial temperature. In practice, however, it is difficult to quick freeze items more than 15 cm thick.

As a general rule, in order to obtain high freezer performance, R_s should be less than 0·04 m^2 degC/W and preferably less than 0·02 m^2 degC/W. It can be seen from Fig 8.2 that R_s exerts less influence in the freezing of the larger fish.

Freezer Output

Again, assuming the properties of ordinary wet fish, the refrigeration requirement is given in Fig 1.2. Although this graph strictly applies to white fish, the values will be only slightly higher than those for fatty fish and so can be used generally for the estimation of refrigeration load. Taking the initial temperature as 0 °C and the final temperature as −30 °C, the freezer must extract 300 MJ/t. If the initial fish temperature is higher, the freezer also will have to extract some sensible heat in the region above 0 °C. The additional refrigeration load will be 40 MJ/t at an initial temperature of 10 °C and 80 MJ/t at 20 °C.

Thus the refrigeration capacity required for a given output is easily estimated. Allowing for an initial temperature of 20°C, the capacity required for a design output of 1 t/h is 105 kJ/s. Additional loads imposed by the cold store and other factors may give a total requirement of more than 140 kJ/s in such a case, but the freezing load will be by far the greatest.

There are losses due to dehydration in all methods of freezing. The amount depends on the type of fish, wrapping, freezing conditions, etc., but it is usually less than 2 per cent with good practice. High losses, when they occur, usually can be attributed to poor cold storage following freezing. The loss in market value due to dehydration is often in direct proportion to the loss in weight and can be a factor in the choice of freezing method.

Automatic freezers of various designs are under development and may be introduced in the future, especially continuous freezers for processed fish in regular shapes, with the product in direct contact with the metal evaporator. An example is the continuous rotary freezer for shrimp and other fish products. It consists of a horizontal refrigerated cylinder with special conveyors for feed and discharge. The fish remain in contact with the cylinder for nearly one revolution, then they are knocked off by a stationary bar. The following discussion of freezers, however, has been confined mostly to fundamental considerations and the main types of freezers in use; air blast, plate and immersion freezers.

Output and freezing time

The output of a freezer will depend on the freezing time and the amount of fish in the freezer. This principle applies to batch and continuous freezers. In the two examples below, the freezer cycle includes the freezing time plus any allowance for loading and unloading that might be necessary. The refrigeration capacity is about 150 kJ/s for a desired output of 1½ t/h.

Freezer 1 (for large fish or items)

freezer cycle	6 h
fish in freezer	9 t
output, 9/6 =	1½ t/h

Freezer 2 (for small fish or items)

freezer cycle	2 h
fish in freezer	3 t
output, 3/2 =	1½ t/h

Two situations are of interest here; the freezing of large items in Freezer 2 and the freezing of small items in Freezer 1.

Assuming that Freezer 2 will hold only 3 t of large fish, the output will be reduced to $\frac{1}{2}$ t/h with the longer freezer cycle of 6 h. The freezer is too small for the desired output with large items and the refrigeration plant will be under light load.

If 9 t of small items are loaded into Freezer 1, the output cannot be increased to $9/2=4\frac{1}{2}$ t/h because the refrigeration capacity is limited. An attempt to exceed the normal output of $1\frac{1}{2}$ t/h will result in incomplete freezing. Either the freezer should be partly filled or the freezer cycle should be lengthened when smaller items are frozen.

Constant freezing conditions such as refrigerant temperature have been assumed in these examples. In fact the simple calculations will be only roughly correct because refrigerant temperature and refrigeration capacity can vary to some extent according to the load. The points illustrated in the examples apply nevertheless and must be borne in mind when choosing and operating freezers. This is particularly so for air blast freezers used for the freezing of a variety of items with different freezing times. Also, when there is simultaneous freezing of various items with different freezing times, care must be taken to ensure complete freezing of all of them.

Air Blast Freezers

The principle of the air blast freezer is shown diagrammatically in Fig 8.3. Air is recirculated by the fan through a heat exchanger or cooler and over the produce to be frozen. The produce gives up its heat to the air which in turn is cooled by the heat exchanger. The main advantage of the air blast freezer is its suitability for all shapes and sizes of product. Its main disadvantages are that freezing times tend to be longer and it occupies more space and requires more power than other types, mainly because of the relatively poor heat transfer properties of air. The design and operation of the freezer on board ship are along the same lines as shore installations except that the motion of the ship must be taken into account.

The air blast freezer is satisfactory from the point of view of fish quality. There is no contamination of the fish. In some cases where distortion of fish frozen in the plate freezer is considered to be a disadvantage, the air blast freezer may be preferred, particularly for whole fish which are to be frozen singly and then sold in the frozen state. Distortion can be reduced or eliminated by air blast freezing where there is little or no pressure exerted on the fish, especially when they are hung.

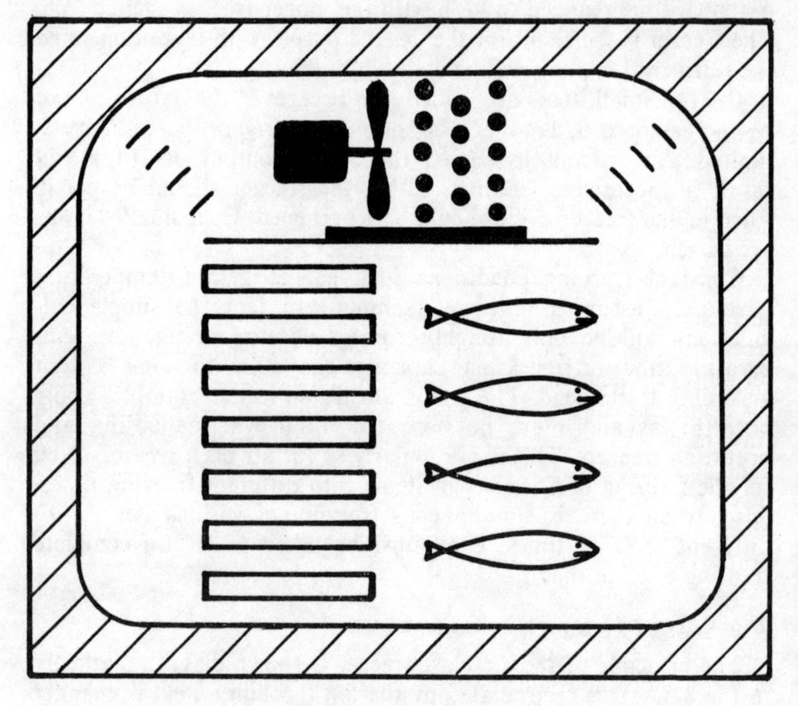

FIG 8.3 Air blast freezer

The freezing rate mainly depends on

- the air temperature,
- the air velocity,
- the product.

Evaporation of moisture from the fish during freezing, although undesirable, also can have some influence. Just as moisture can migrate from produce to cooler in the cold store, as described in Chapter 1, so there can be a transfer of moisture on freezing. Since the latent heat of evaporation of moisture is about seven times the latent heat of freezing, a small amount of evaporation can have a significant effect, reducing the surface resistance, R_s, and increasing the freezing rate. Weight loss through evaporation will tend to be highest in small items frozen singly. In quick freezing, however, weight losses are rarely over 2 per cent in a freezer of good design except in very small items and there is no significant quality damage.

The product

Although transfer or moisture can have some effect, given the air temperature and velocity, the freezing rate depends by and large on the nature of the product. Before the design or selection of a freezer, the types of products and freezing times should be known. In simple cases the freezing time can be estimated with the help of Fig 8.2 or by judgement based on experience but it is usually desirable to determine the freezing time by measurement if it is not known accurately.

Wrappers and containers may be looked upon as increasing surface resistance, R_s, increasing the freezing time and reducing freezer output. Often packing is easier and the packs more compact for storage when wrapping is done before freezing. In some cases, for example in the freezing of blocks of fillets, the fish is placed in trays or moulds before freezing and then removed after freezing for packing and storage. When choosing wrappers, containers and trays, the question of freezer output should be considered. Their effect can be considerable and they actually can increase dehydration over the amount obtained with unwrapped fish, especially if there are spaces of trapped air which provide a good deal of insulation. Thus while a tight container of wood 5 mm thick would add about 0·03 m^2 degC/W to the resistance, a container with a lot of air space would add more. Even an aluminium container could have a considerable effect in this way.

There are instances where a tray or similar support can reduce the freezing time. A single fish will freeze slightly faster on a large sheet of aluminium in air blast than one suspended by itself.

Where metal trays with lids are used, the degree of contact between lid and fish can have a significant effect on freezing time, as in the horizontal plate freezer. Hence there has been an interest in lids that are clamped on to the tray in order to improve contact. In one design of air blast freezer with trays of cast aluminium, this type of lid is used along with extended surfaces in the form of fins on the bottom of the tray and on the lid, giving freezing times approaching those obtained in the plate freezer. The fins run more or less parallel to the direction of air flow and are far enough apart to realize a high air velocity between them.

Air temperature

Air temperatures are normally in the range -35 to $-40^\circ C$ for freezing, to be followed by storage at $-30^\circ C$.

Air velocity

Since air is a poor conductor of heat, in order to carry away heat from the surface of the fish at an appreciable rate it is necessary to have air movement or convection. Forced convection by means of fans is used because natural convection is insufficient for high heat transfer rate. For the same reason, forced convection is used at heat exchangers such as the cooler in the air blast freezer. The freezing rate, hence freezer output, can be increased by increased air velocity. Successive increases in velocity yield diminishing returns, however, and in any event the resistance of the fish itself determines the maximum possible cooling rate. These points are illustrated by data given in Table 1.1 and in the above discussion of rate of freezing.

The fan usually is driven by an electric motor which should be powerful enough to maintain the desired flow rate of air against the resistance offered by the cooler, produce, freezing chamber and ducting. The power required increases as the cube of the velocity, or nearly so. Since most of the power is converted to heat in the air, the cooling load is increased by increased velocity. The required refrigeration capacity must include an allowance for fan heat.

Because of these factors there is a practical limit to the velocities that should be used in air blast freezing. Normally the air velocity should be in the range 5 to 7 m/s which can give surface resistances in the region of 0·04 to 0·03 m^2 degC/W. Higher velocities have been used in some installations but they can hardly be justified except where small products such as fillets and some shellfish are frozen individually in a continuous freezer. Sometimes lower velocities are used, particularly for large fish.

It should be noted that it is the velocity over the product which must be specified. Too often the distribution of air is poor so that freezing rate will tend to vary from place to place in the freezer. The design of the freezer should ensure uniform velocities. Baffles can be installed and adjusted, with the aid of a suitable instrument for the measurement of velocity, to give more or less even velocity. When the performance is checked by measurement of freezing time, care should be taken not to overlook any variations in freezing time due to variations in velocity. If output is based on an optimistic forecast of freezing time, some of the fish will be incompletely frozen with bad effect on quality.

The mean velocity and the velocity distribution in the freezer will be affected by the size, number and disposition of the product. These points must be borne in mind by the operator. Overfilling of the freezer can completely prevent air circulation in an extreme case. What is more likely, however, is uneven disposition of the product so that, although the velocity distribution may be uniform in the empty chamber, the velocities when loaded and hence the freezing times are uneven. The produce should be evenly spaced, as indicated in Fig 8.3, with full loads and with part loads. There should be no large open area through which the air can short circuit. As a general rule, the produce should not occupy more than ½ the cross-section. Thus the working velocity will approach double the velocity in the empty freezer, depending slightly on the fan characteristics.

Fans

Centrifugal and axial fans can be used but usually the latter type is preferred because of its inherently higher capacity and because the resistance to air flow in the freezer normally is low, consisting mainly of the resistance of the heat exchanger. Very often a simple propeller fan is mounted directly on the heat exchanger so as to discharge air through it.

The power required will depend on the capacity and resistance to flow. With good design, the fan power will be substantially less than ½ the net cooling load, that is less than 50 kW for a rated output of 1 t/h (approximately 100 kJ/s cooling load). Often fan power is less than ¼ the net cooling load, less than 25 kW per 1 t/h of rated capacity.

The mass flow rate of air (kg/s) should be sufficient to carry away the heat from the produce without too great an increase in air temperature. For example, if the permissible temperature rise across the freezing chamber is 1 degC, the fan should circulate approximately 100 kg/s (67 m^3/s at $-40\,^\circ$C) for each 1 t/h output of frozen fish. If the permissible rise is 2 degC, the fan should circulate only 50 kg/s (33 m^3/s) for each 1 t/h, and so on.

Heat exchanger

The heat exchanger or cooler usually consists of a compact unit of not more than six rows of finned metal tubes, operated on a dry expansion, flooded or pump circulation refrigeration system. Often the cooler and fan are manufactured as one unit with a net rated cooling capacity. In addition to the heat from the fan, the cooler must cope with the heat from the fish, by far the largest cooling load, and any heat introduced by wrappers, trolleys, etc., and by conduction and air changes which usually are relatively small amounts. The required cooler surface depends on a number of factors including air velocity and refrigerant conditions. Most of the resistance to heat transfer is on the air side, hence the use of extended surfaces in the form of fins. The design temperature difference between the air and the refrigerant is typically 5 degC. Thus, as in any system with a secondary refrigerant, air in this case, there is the disadvantage of relatively low suction temperature.

In some installations the heat exchanger consists of coils of plain pipe around the inside of the freezing chamber, particularly where modest velocities are employed for large fish. Ideally, however, the fan should discharge through the cooler, not directly through the freezing chamber. Since the cooler is the coldest part of the air blast system, any moisture picked up by the circulated air, usually from the fish and from the introduction of warm outside air into the system, tends to be deposited on it as frost. The air leaving the cooler will be saturated with water vapour and fan heat added after the cooler can increase the amount of evaporation from the fish. This undesirable evaporation will manifest itself as frost on the cooler, increasing the need for defrost.

The amount of cooler surface required depends mainly on air velocity with some allowance for frost depending on the configuration. Usually the velocity over the surface does not exceed 7·0 m/s. At the normal recommended temperature difference of 5 degC between air and refrigerant, the amount of surface required will be roughly

0·8 dm^2/W for 7·0 m/s
1·2 dm^2/W for 3·5 m/s
2·4 dm^2/W for natural convection

Defrost

In order to maintain high performance, the cooler will have to be defrosted periodically, the frequency depending on the conditions of operation. In a finned cooler, the fins should be spaced far enough apart to prevent their becoming quickly bridged by frost, making them ineffective. In some coolers the distance between fins is varied, with the wider spacing, or perhaps no fins at all, at the upstream tubes where the frost tends to be heaviest. More than one cooler may be installed in an air blast freezer in order to give good velocity and temperature distribution and possibly permit the defrosting of one cooler at a time without having to interrupt the freezing operations.

Usually the most suitable methods of defrost are hot gas and electric heating. They can be initiated manually or automatically. Frost can be brushed off a plain pipe. Natural defrosting by turning off the supply of refrigerant and opening the doors of the freezer so that warm air can circulate through is usually unsuitable. The interval between defrosts should be short enough to ensure that there is no serious reduction in heat transfer and air flow at the cooler, and that the accumulation of frost is light enough to be removed easily.

The freezing chamber

The freezer should be of such a size and so arranged that the requirements of low fan power and high air velocity are reconciled. The load of produce in the freezing chamber should impose only a small resistance to the flow of air. Again taking the 1 t/h unit with a rise of 2 degC as an example, the fan capacity will be 33 m^3/s. In an empty freezing chamber with a cross section of 10 m^2 the average velocity will be 3·3 m/s and when the chamber is loaded the average velocity over the produce will be not more than 6·7 m/s, assuming that the free area is reduced to not less than 5 m^2 by the load.

The freezing chamber, and preferably the entire freezer, should have a watertight lining of sheet metal or some other suitable material in order to protect the insulation from moisture and provide an easily cleaned inside surface. Any water introduced or meltwater produced by warming of the freezer when it is shut down should drain away and not remain where it will be frozen.

Loading and unloading should be as straightforward as possible with no danger to the crew. Particular attention should be paid to access doors, since heavy doors can be dangerous at sea if they are not under proper control. Heaters should be installed to prevent doors from freezing shut. Curtains have been used in place of doors in some installations, mainly on grounds of convenience and safety.

The freezer itself can take one of a number of forms broadly classified as batch, semi-continuous and continuous.

Insulation and vapour seal

The air blast freezer should have insulation and vapour seal along the same lines as the cold store, discussed in Chapter 9. The entire air blast freezer should be insulated in order to prevent the entry of a large amount of heat from outside. The thermal resistance need not be as high as in the cold store but usually the insulation thickness is 5 cm or more, giving a resistance of at least 1·4 m^2 degC/W.

The vapour seal should be complete in order to protect the insulation against the ingress of water vapour and the whole structure should be tight so that exchange of air between outside and inside is held to a minimum. A high rate of air change implies increased cooling load and accumulation of frost, at the coolers and wherever outside air enters.

Batch freezer

In a batch freezer the produce may be on shelves or trolleys or suspended from hooks. Trolleys are convenient because some of them can be loaded while others are in the freezer. The design of trolley should be such that correct air distribution is achieved, with no short circuits for air between the trolley and the sides of the freezing chamber. For safety reasons trolleys are best operated on a track and equipped with a system of brakes so that the motion of the ship will not create difficulties.

Where there are two or more trolleys or a number of sections in the freezer, the flow of air may be across them, cross flow, or along them one after the other, series flow. With cross flow there usually is more than one cooler and fan unit required, but this can have an advantage for defrost.

A disadvantage of the batch freezer is that the cooling load is uneven, as indicated in Fig 8.1. This can be compensated to some extent by the operation of several freezers in sequence. Its greatest advantage is that of flexibility. Freezing periods can be adjusted easily to suit the product.

A separate small batch freezer often is employed for large fish when they make up the lesser part of the catch. Since low surface resistance is not so important for large fish, a modest air velocity is used, perhaps 2 m/s or so. In some vessels such a freezer is essential. There is sometimes a tendency to leave the fish in the freezer too long and use the freezer as a store. This can interfere with air circulation and the freezing of fish as well as promote dehydration.

Semi-continuous freezer

In a semi-continuous freezer the freezer load is divided into several parts. Periodically, depending on the freezing time, one part is withdrawn from one end of the freezer and a new part put in at the other end. This is done with the trolley system, for example where a train of trolleys passes through the freezer. Both cross flow and series flow are employed. A relatively high temperature rise can be tolerated with series flow, which will reduce the required fan capacity. Cross flow has an advantage on defrost where several coolers are employed but cooling load and frost accumulation are uneven, higher at the end at which the fish enters. The semi-continuous freezer can reduce the variations in compressor load, however, and can smooth the flow of frozen fish to the cold store.

Continuous freezer

Continuous freezers equipped with various types of belt conveyors are used in factories ashore but generally not at sea. They are well suited to the freezing of a variety of small retail products. The more efficient designs enable rapid freezing at high output for periods of more than 20 h before shutdown for defrost is required.

Continuous air blast freezers have been developed, however, for freezing at sea. One example is an automatic freezer for large tuna in which the tuna are hung on an overhead conveyor and pass through the freezing chamber. The conveyor can be driven continuously but also can travel intermittently as in the semi-continuous freezer. Another example is a cross flow type of freezer with a conveyor made up of a series of finned aluminium trays about 8 cm deep for smaller fish. The trays are heated as they leave the freezer and the lids are removed, releasing the blocks of fish which are conveyed to the cold store, all done automatically.

PLATE FREEZERS

In the plate freezer, sometimes called a contact freezer, the product is placed directly in contact with refrigerated metal plates. Basically there are two types, the horizontal plate freezer and the vertical plate freezer. Given a good refrigeration system and a high degree of contact between fish and plate, the effective mean surface resistance, R_s, will be less than $0{\cdot}02\ m^2$ degC/W. Thus freezing times tend to be relatively short, giving a compact freezer that is simple and robust. As in air blast freezing, wrappers and containers can increase the resistance, of course, adding $0{\cdot}01\ m^2$ degC/W in a typical case. Trapped air spaces in a pack will encourage dehydration.

The plate freezer does not have the versatility of the air blast freezer and sometimes deformation caused to the fish by the plates is considered to be a disadvantage. Plate freezer systems require a good deal less power than air blast systems, however, mainly because of the power requirements of the fan(s) in the air blast freezer and because the temperature difference between air and refrigerant at the cooler implies a lower suction temperature in the air blast system.

It is recommended practice to install a small air blast freezer for large fish in vessels which are equipped to freeze most of the catch in plate freezers.

The horizontal plate freezer

In the horizontal plate freezer there are a number of plates one above the other, perhaps more than fifteen with a total area in excess of 150 m^2, with a space of usually not more than 8 cm between them. The space is adjustable, usually by means of a hydraulic or pneumatic ram which moves the plates up and down with one fixed plate at the bottom. There are flexible connections in the piping to the plates. Loading, by inserting the produce to be frozen between the plates, and unloading are done with the plates set at the open position. Both top and bottom of the product can be brought into contact with the plates before freezing and some compression, typically 50 kN/m^2, may be exerted in order to reduce freezing time as far as possible by high contact and, if desired, compact the fish. The compression system also must allow the freezer plates to move apart under the force exerted by the expansion of the fish on freezing, otherwise the plates will be damaged. Care should be exercised in loading the plate freezer, particularly at part loads, so as to prevent distortion of the plates through uneven loading. The force of compression should be taken up more or less evenly, if necessary with the help of removable spacers of wood or some other suitable material.

The plate assembly normally is enclosed in an insulated cabinet. An insulation thickness of 5 cm giving a thermal reistance of more than 1·4 m^2 degC/W is usually more than enough. The main function of the cabinet is to prevent excessive frost formation on the plates through contact with warm, humid air. Relatively light cabinets with lower thermal resistance have been installed in some cases in order to reduce first cost and save space on board. Curtains have been introduced in place of hinged doors and have much to commend them, particularly from the point of view of safety on board ship.

Loading of trays or packs into the freezer can be awkward or even dangerous in rough seas. Automatic loading systems and semi-continuous methods in which the product is intermittently pushed through the space between the plates, entering at one side through a small opening in the cabinet and leaving at the opposite side, have been developed but are not used widely on fishing vessels. Batch loading is almost always employed, but it is usually good policy to have four or more freezers, operated in sequence, rather than one or two larger freezers in order to even out the cooling load.

The horizontal plate freezer is particularly suitable where the produce is to be frozen in trays or in packs of a regular rectangular shape. It is used for the freezing of blocks of fillets and for whole fish frozen individually and in blocks. Fillet blocks for the production of fish sticks ashore should be free of voids and have square edges, hence they are frozen in metal trays or moulds under pressure in the freezer in order to compact the block. They are pressed from the moulds or released with the aid of heat on removal from the freezer before stowage.

With some types of pack, for example with boxes that are incompletely filled, there is poor contact at the top. Consequently freezing time is longer and output is relatively low. There may be some danger of incomplete freezing. The loading of packs of different thicknesses can have the same effect because only the thickest packs will be in contact at the top. Even with only slight differences in thickness, the more rapid freezing of the pack or tray of fish in good contact may force the plate further away from the other packs. Most often, however, poor and uneven contact is caused by ridges and nodules of ice which have built up on the plate. These accumulations of ice form at the edges of trays or between packs. An efficient defrost system is essential to maintain high output and prevent difficulties in loading.

The vertical plate freezer

In many respects the vertical plate feezer is similar in operation to the horizontal plate freezer. Unlike the horizontal plate freezer, however, it was developed primarily for the freezing of whole fish at sea on distant water vessels in order to supply processing factories ashore. It has been used mainly for the freezing of whole fish such as cod, haddock, herring, redfish and hake and to a limited extent for the production of frozen fillet blocks and some packaged products.

The product is put into the freezer through the openings at the top. Provided the plates are not refrigerated until loading has taken place, loose items such as whole fish can settle between the plates by virtue of their own weight, making a compact block with good contact between fish and plate. Usually contact is better and freezing times are shorter than in the horizontal plate freezer where contact between plate and top of the fish is not as good. Loading of the vertical plate freezer is easier because it is assisted by gravity and trays are not required. Loading has been done manually and, in a few cases, from hoppers.

The typical block of whole fish such as cod is 100 cm long by

50 cm high by 10 cm thick, 50 dm^3 in volume, weighing 40 kg. This gives a compact block which can withstand a reasonable amount of handling without breakage. Sometimes thinner blocks are produced when freezing relatively small fish, particularly species with soft flesh that may not tolerate freezing as well as firmer fish. Water may be used to fill the spaces in the block in such cases. Sprats have been frozen in special watertight freezers in this way, largely to provide good protection against physical damage after freezing. There also will be some protection against dehydration during storage and possibly some benefit from reduction in stress during freezing. In some cases watertight wrappers, placed in the freezer compartments immediately before filling with fish and water, have been used. They enable the addition of water in ordinary freezers and protect the fish during handling and cold storage. Storage life can be doubled by this technique. Plastic wrappers are not always suitable because they provide little friction between frozen blocks and therefore there may be danger from unstable cargo. Paper with a plastic coating on the inside has been used and has good frictional properties but is easily torn during handling. Such material has to be thin, otherwise freezer output will be reduced. It is normal practice to place cartons or wrappers in the freezer spaces before filling with fillets. The block of fillets usually is not more than 6 cm thick.

The freezing time of the block 10 cm thick can be less than 3 h under ideal conditions with a refrigerant temperature of $-40\,^{\circ}C$. This implies a mean value of surface resistance, R_s, of not much more than 0·01 m^2 degC/W as indicated in Fig 8.2. In practice, the time allowed in the freezer is 3 to 3½ h for the block of 10 cm.

An insulated cabinet usually is included, sometimes with a lid. As in the horizontal plate freezer, the plates should exert some pressure against the fish in order to maintain good contact but they should be able to take up the expansion of the fish on freezing. This has been done by means of springs in some designs but the hydraulic system is the more common method. In any case there also must be some form of flexible connection in the refrigerant piping to the plates. The plates are closed against members of the appropriate dimensions installed near the edges of the plates. As well as providing stops, these members contain the load and may form part of the mechanism for unloading. In some types with hydraulic systems the distance between plates can be changed without a great deal of trouble, if a change in block thickness is desired, by changing the members.

With the hydraulic system, the plates are moved apart for ease of unloading and loading, increasing the distance between plates by about 12 mm. The plates are returned to the closed position after loading but before freezing begins. With the plates open at the top, this may give some compaction of the block and improved contact with the softer fish but has only a small effect with firm fish such as fresh cod in good condition. The degree of compaction can be greater where a lid is closed before the plates are moved to the closed position, a procedure which may lead to some physical damage to the fish. In one type of freezer, specially designed for the freezing of solid blocks of fillets of accurate dimensions for the production of fish sticks or portions, the hydraulic system exerts a pressure at the top of the block through movable members or 'forks' which fit between the plates.

There are a number of methods of unloading. Hydraulic systems have been used to discharge at the top, through a door at the end and by lowering the fish out of the bottom of the freezer on a platform. In some types the blocks are dropped through doors at the bottom of the freezer. In one model without hydraulics the block of fish is pushed out of the end by a lever which the fisherman operates with his foot. Top unloading is convenient where the space between plates is divided on loading in order to produce more than one block or pack of reduced size and where more care than usual must be exercised in loading, since the base members or 'forks' can be set at any level between the top and bottom of the plates. With bottom unloading the blocks can be discharged directly through the deck into the cold store. This has the advantages of saving space and labour in some cases and eliminates the possibility of heating of the frozen fish before stowage.

Defrost

Defrosting of the plate freezer at regular intervals is essential if high performance and ease of operation are to be maintained. Defrosting can be carried out by scraping the plates and by natural defrosting, subjecting the freezer to warm ambient conditions, but these methods are not recommended for freezing at sea. The internal hot gas defrost, or hot defrost when a secondary refrigeration system is used, is the best method. The normal fluid temperature for defrost is approximately 33°C and the defrost period is usually short, 1 to 3 minutes. When carrying out the defrost operation, care must be taken not to heat the product unduly, both before and after it is frozen.

The vertical plate freezer is a special case. With whole fish and some other products, a short period of defrost immediately after the freezing period releases the block of fish and makes unloading easier. Even more important, however, is the effect of defrost on the loading of the freezer. The product should be removed from the freezer as soon as it is released by the defrost and the defrost period should be as short as possible so as to avoid heating of the next load. Loading should be carried out with the supply of hot fluid shut off but also before freezing begins. By loading with the temperature of the plates above the freezing point of the fish, the blocks will be compact and output high. If the freezer is loaded after the low temperature refrigeration has been turned on, the fish will tend to stick to the plates. Observations on the freezing of whole cod in blocks 10 cm thick have shown that lack of a defrost or incorrect procedure can result in easily broken blocks of more than 20 per cent lower density and freezing times up to 1 h longer. Thus the freezer output is reduced to not much more than half the possible output and the stowage rate in the cold store is reduced by at least 20 per cent.

Refrigerated plates

Ideally the freezer plate should have a flat, smooth surface, it should be robust and impose a negligible thermal resistance between the product and the fluid in the plate so that freezing and defrosting are carried out as efficiently as possible. Resistance to the flow of fluid should be low. There was a marked improvement in the plates used in commercial freezers, beginning about 1960. The later plates approach the ideal whereas the earlier types were poorly designed. Freezing times were longer and response on defrost was slow, so that release of the block in the vertical plate freezer could not be accomplished quickly enough to avoid overheating of the fish.

High performance has been obtained using plates made from hollow sections of extruded aluminium. These sections are typically about 3 cm thick by 25 cm wide by 1 m or more in length with several rectangular ducts running lengthwise, each about 4 cm^2. Similar high performance has been achieved with plates of cast aluminium with drilled passages and with plates fabricated out of sheet steel.

The ducts usually are connected in series by suitable end sections although more than one circuit may be used in the larger plates. Pump circulation systems with primary refrigerant are preferred to dry expansion because of their simplicity and better heat transfer. According to one method, the expansion of the refrigerant as it evaporates is accommodated by connecting some of the ducts in parallel, the number of parallel ducts increasing as the refrigerant passes through the plate. In this way the resistance to flow and difference in refrigerant temperature between inlet and outlet due to pressure drop is kept low.

The value of surface resistance in the duct is low, if the values of pumping rate given in Chapter 7 are employed. With the secondary refrigerant trichloroethylene at $-45\,^{\circ}C$ the velocity in the duct usually is more than 30 cm/s. Thus the resistance will be less than 0·003 m^2 degC/W (heat transfer coefficient 333 W/m^2 degC) which is entirely satisfactory for most applications. The value tends to be even lower in the primary system and, for all practical purposes, the thermal resistance of the metal plate itself will be nil.

Flexible connections

The flexible connections in the refrigerant piping, to allow movement of the freezer plates, often have been a source of trouble because they are prone to leak. Leakage can be unsafe for the crew, contaminate the product in the freezers and interfere with the proper functioning of the refrigeration system through loss of refrigerant and ingress of air. It also can be costly in terms of loss of refrigerant, particularly primary refrigerant. In the selection of flexible connections, the emphasis has to be placed on reliability. They must be compatible with the refrigerant and any oil that may be present in the refrigerant.

O-ring seals have been employed in some horizontal and vertical plate freezers with limited success. In some cases there have been difficulties created by the fluctuations in temperature and pressure encountered during the normal operations of freeze, defrost and shutdown. In one type of vertical plate freezer with only a small amount of movement to allow for the expansion of the fish, coiled copper pipes have bcen used. The most common type of connection, however, is the flexible hose with an internal diameter of 10 to 30 mm and up to more than 1 m in length, depending on the type of freezer. Various kinds of hoses have been used, including reinforced rubber and plastic (polytetrafluorethylene) tube with outer stainless steel braid. Hoses of stainless steel have been coming into wider use because of their reliability but their initial cost has been high.

Combined plate and air blast freezer

The combined horizontal plate and air blast freezer has been used in a few installations on shore and on board ship. Essentially it is an air blast freezer in which the produce, perhaps on trays, is placed on shelves of refrigerated metal plates. Its heat transfer performance can be somewhat better than the horizontal plate freezer in some cases where the contact between the top of the product and the plate above is poor or impossible. Thus for a given output the combined freezer may be smaller in size and consume less power than the air blast freezer but retain some of the advantages of the air blast. It is loaded in the same way as the horizontal plate freezer, which is a disadvantage for most freezing at sea applications.

Large fish have been frozen in direct contact with refrigerated pipe coil in chambers with modest air circulation.

Immersion and Spray Freezers

Freezing by immersion of the fish in a cold liquid or by spraying is sometimes used because of its inherent simplicity and high output. In comparison with air, liquids have much higher thermal capacities and thermal conductivities. Heat transfer can be rapid even with moderate circulation because there is good contact with the product and, unless the viscosity of the liquid is high, the thermal resistance at the surface of the fish is low. Thus, like the air blast freezer, the immersion freezer will accommodate fish of various sizes and shapes but it will be more compact and consume less power. Both batch and continuous immersion freezers have been employed for the freezing of poultry, fish and other foods in shore installations. The method has found only limited application for fish, however, its use at sea having been confined mainly to the tuna fishery of the Pacific and to a lesser extent for the freezing of large halibut and salmon, where most of the material is destined for canning ashore. Probably the main factors which have prevented more widespread use of the method for the freezing of other species including shellfish are the contamination of the fish by the fluid and difficulty in handling and storage. The bulk density of frozen whole fish, stored as single fish, usually is less than 500 kg/m^3.

Contamination can be eliminated by the use of wrappers but this would not be contemplated for freezing at sea in most cases because of the added difficulties in handling and increase in freezing time. Therefore the number of suitable fluids is restricted. A mixture of water, sugar and alcohol could be used and a solution of propylene glycol and water has been proposed. Immersion in Refrigerant 12, with direct evaporation of the refrigerant at atmospheric pressure, temperature $-30^\circ C$, has been employed in freezers on shore but not at sea. Refrigerated brine made from a solution of common salt, sodium chloride, in water is of the most practical interest as it is used in the tuna fishery. With all these liquids there is the problem of reduced acceptability of the fish due to contamination.

Sodium chloride brine

Sodium chloride brine can be made up to any desired strength according to the desired freezing point of the brine. A saturated solution, which contains about 26 per cent salt, freezes at $-21\,^\circ C$. In practice, therefore, the operating temperature cannot be lower than about $-18\,^\circ C$. This is a serious limitation because the refrigerant temperature should be below $-30\,^\circ C$ and the storage temperature should be $-30\,^\circ C$ or lower in most applications. Even with the brine at $-18\,^\circ C$, however, it is possible to achieve rapid freezing. The surface resistance with immersion will be less than $0{\cdot}02\ m^2$ degC/W with mild agitation. Based on a temperature rise of the brine of 2·5 degC, the circulation or pumping rate should be about $10\ dm^3/s$ for a freezer output of 1 t/h (approximately 100 kJ/s).

The amount of salt taken up by the fish depends on a number of factors; increasing with increase in temperature and concentration of the brine and increase in surface area of the fish. Smaller fish will absorb more salt. Generally speaking, for a fish of a given size, the longer the freezing time the greater the amount of salt taken up. Usually there also will be an appreciable amount of salt on the surface of the frozen fish after it has been removed from the brine.

As mentioned in Chapter 1, the presence of salt can speed up some of the changes that take place in cold storage. The storage of single fish, as opposed to compact blocks, also may lead to increased dehydration, oxidation and other changes reducing the storage life. A reduction in storage life may not be crucial, however, and a small amount of additional salt is not a disadvantage in some cases such as in fish for canning.

Ultra-rapid freezing

From time to time proposals have been put forward for the freezing of fish and other foods by direct contact with liquified gases which evaporate at temperatures well below $-80\,^\circ C$ at atmospheric pressure. There have been a limited number of applications on shore but none on fishing vessels. The most notable of these liquid gases are liquid nitrogen, liquid air, liquid nitrous oxide and liquid carbon dioxide. Heat transfer conditions closely approach the ideal, giving rapid freezing and hence a very compact freezer. There is little or no problem of contamination.

Since about 1960 interest has increased, particularly in liquid nitrogen and liquid air. Liquid nitrogen has been used in refrigerated transport and in freezers for individual fillets and for shrimp which are sold for a relatively high price.

In freezing, the rate of evaporation of nitrogen, at −196 °C, will be not less than 1 kg per kg of fish. In addition there may be the requirement for the cold store and other small losses. Thus the generation of liquid nitrogen on board has been proposed, as opposed to the system used for refrigerated transport in which the comparatively small amount of liquid nitrogen required is held in an insulated container. The inclusion of some storage capacity for liquid nitrogen in a generation system is an advantage, however, where there are fluctuations in demand. Plant suitable for the production and storage of liquid nitrogen on board and the freezer themselves are available. The cost of producing liquid nitrogen is high compared with the conventional methods of refrigeration at −40 °C, however, mainly because of the fundamental rule that power requirements are increased by a reduction in temperature. The increase in power in this case is roughly ten times. This disadvantage in cost is not offset to any extent by advantages in quality over the normal methods of quick freezing except that weight losses will be lower than in air blast freezing.

In practice, the freezing rate employed in a liquid nitrogen system for unwrapped fish is only about $\frac{1}{4}$ the maximum rate possible by direct immersion. Contact between the liquid nitrogen and the fish is restricted by control of the supply of nitrogen, usually through sprays. This is done for two reasons; to utilize the cooling capacity of the cold nitrogen vapour and to avoid damage to the fish due to thermal stresses. Much of the cooling of the fish, in the initial stages, can be done by the cold nitrogen vapour. Heating of the vapour to −20 °C by the fish accounts for roughly half of the refrigeration requirement, with the other half to be supplied by the evaporation of nitrogen. Thus the rate of evaporation can be limited to not much more than 1 kg per kg of fish.

The thermal stresses set up in the fish while freezing can be considerable and can cause fracture of the flesh at low temperatures. Evidently there are two mechanisms; one having to do with the expansion of the material by about 8 per cent on freezing and the other with differential rates of contraction when the frozen material is cooled rapidly. Since the outside of the fish forms a shell before the inside becomes frozen, the outer shell will be under tension and the inside under compression when the inside expands on freezing. Rapid cooling of the frozen material apparently makes a lesser contribution. Fracture of the fish does not occur at the conventional rates of freezing, probably because the frozen material is sufficiently plastic to yield at the higher temperatures, but it becomes harder and more brittle as the temperature is reduced. If fracture of the flesh is to be avoided, indications are that the freezing time cannot be reduced much below the value given by the curve $R_s = 0$ in Fig 8.2 with an applied temperature of $-50^\circ C$.

Chapter 9

The Cold Store

COLD STORE DESIGN AND OPERATION

SINCE fish is a highly perishable foodstuff, the design and operation of the cold store are very important. Emphasis should be placed on the temperature requirement of -30°C or below for fish and the features which reduce dehydration of the fish in the store, bearing in mind that these are the two most important considerations in the prevention of protein changes and oxidation, as outlined in Chapter 1. Failure to pay attention to these two points probably has been the most common cause of poor quality where it has occurred.

In many respects the design and operation of the cold store are similar to the design and operation of the insulated wet fishroom which is discussed in Chapters 3, 4 and 5. The most important difference in construction is on the sealing of the insulation against the ingress of moisture. It is essential in cold store construction to seal the warm side of the insulation against the entry of moisture.

The method of stowing the frozen fish varies. Whole fish often are stowed singly or in blocks without any wrapping or container. Blocks of fillets and other processed fish usually are wrapped and packed in cartons before stowage, sometimes before freezing. Pallets and containers of wire mesh have been used on board, mainly for ease of unloading. Automatic conveyors also have been used for unloading from ship, sometimes passing directly into the cold store on shore, avoiding the possibility of undue increase in temperature on transfer.

Heat gains and moisture migration

Heat can enter the cold store in various ways as follows:

the fish	pipes, stanchions, etc.
electric apparatus	bottom, sides and deckhead
the fisherman	bulkheads
air changes	

The first, entry of heat by way of the fish, should be strictly avoided. The other ways of gaining heat are discussed in Chapter 3.

The allowance for the fisherman will have to be higher than in the wet fishroom because of the lower temperature, about 0·40 kJ/s.

Since the cooler is the coldest part of the store, excess water vapour in the air will condense on it as frost, as in the air blast freezer. The moisture can come from various sources; a large fraction of the heat generated by men working in the store will be in the form of latent heat (vapour), ambient air entering the store through hatches and any other openings will carry substantial quantities of water vapour, and moisture may be evaporated from the fish itself. A high rate of transfer of moisture to the cooler should be avoided because it imposes a load on the refrigeration plant and may make more frequent defrosting necessary. Doors or hatches in the store should be open for as short a time as possible and, as in wet fishrooms, there should not be more than one opening at a time so that the circulation of outside air through the store is held to a minimum. Commonly an access hatch is located at the top of the store in the deck which, although inconvenient in some ways compared with the side door normal in land installations, has the advantage that the cold dense air cannot spill out of the store so easily when it is open.

Devices such as automatic sliding doors, air curtains and plastic or rubber curtains are not used often at sea. A small air lock or vestibule sometimes is installed to limit the amount of air change at the entrance. An effective measure is the practice of transferring the frozen material through restricted openings in the deck, directly into the store.

Heat entering through the hull, bulkheads, pipes, etc., should be held to a low level by the use of appropriate insulation. Even so, the heat through the insulation normally will make up by far the greater part of the cooling load.

Dehydration

The transfer of moisture from fish to the cooler is the most serious aspect of moisture migration because of its effect on the fish. The rate is increased by severe fluctuations in store temperature, agitation of the air and low cold store humidity. It is important to remember, however, that heat is required to evaporate moisture, roughly 2200 kJ/kg. If heat from any source is transmitted to the fish, it can produce dehydration or freezer burn as described in Chapter 1. Often it is not the amount of evaporation which creates a problem, but the nature of it, affecting the texture and appearance.

Heat through the insulation will pass through fish placed against it. The correct procedure is to leave a space between the sides of the store and the produce, all around the store, so that heat gained through the insulation does not enter the fish but is transferred to the cooler directly by the air. Incomplete freezing also can have harmful effects by introducing heat into the fishroom. Freezer burn may occur in fish that had been properly frozen but stored in a disadvantageous location, perhaps at the edge of a stack of incompletely frozen fish and near the cooler.

Because of the lower temperature of the cooler there always will be a difference in vapour pressure tending to drive moisture from the fish. With good design and operation, however, the amount will be small and it will be possible to store fish with limited protection by glazing and wrapping or perhaps no protection at all, depending on the product and the circumstances to be encountered on shore. Usually whole fish are not glazed at sea but may be glazed on landing before they are stored ashore. Fresh water should be used for glazing.

Temperature

The temperature should be uniform throughout the store, at −30 °C or below, and it should be steady. Wide fluctuations in fish temperature during storage will encourage freezer burn and some deterioration through alternate freezing and thawing of small quantities of water in the most exposed parts of the fish.

A compact stack of fish at too high a temperature will cool down to store temperature very slowly because of the high thermal resistance of the fish. The bulk of the stack will remain at too high a temperature for weeks, suffering deterioration as a result, even if the cold store temperature is low enough.

It should be borne in mind that the cold store is unsuitable for cooling the fish. Typically, the refrigeration capacity of the cold store, mostly to cope with the heat gained through the insulation, is less than 10 per cent of the capacity of the freezing plant. Thus the stowage of fish at too high a temperature can result in an intolerably high cooling load. The mean temperature of each item to be stored should not be above the desired cold store temperature.

The fish handling system should be planned and operated in such a way that incomplete freezing and the warming up of fish before stowage is avoided. The cold store should not be used for the freezing of fish. Adequate freezing facilities should exist for all the kinds of fish to be stored and there should be proper quality control during the period before freezing, even when fishing is heavy. The practice, sometimes observed, of operating freezers on too short a freezing period with incomplete freezing in order to increase output is unacceptable. The freezing time, the time required to reduce the fish to the required temperature, and the temperature of the fish in the store can be measured as described in Chapter 8.

A close check should be kept on cold store temperature, preferably at several points with the aid of a permanently installed indicator or recorder. The placing of the sensing elements so as to obtain representative readings is largely a matter of judgement. A survey of temperatures throughout the store, bearing in mind that the pattern of stowage will exert some influence, may help. Two desirable locations are the region of the cooler and near the hatch or opening for the store. The elements must not be placed too close to such areas, however, otherwise the reading will give a poor indication of store temperature. Sensing elements placed in the region of a pipe grid should be more than 30 cm away and shielded against damage.

Linings

In some respects, the function of the lining in the cold store is different from the function of the lining in the wet fishroom, discussed in Chapter 3. The cold store is dry and free from the problem of bacterial spoilage. To a great extent the lining merely serves to protect the insulation from physical damage. In ordinary installations the lining need not be watertight, indeed it may be better if it is not; a lining that permits the passage of water vapour will help to keep the insulation dry. Water which may have found its way through imperfections in the vapour seal or lodged behind the lining during periods when the store has been shut down and at high temperature, will be able to dry out, that is, to pass through the lining as vapour when the store is refrigerated.

Complete linings have been made of sheet metal and wood. They may serve to contain foamed in place insulation when it is installed. In some vessels the lining consists of an open mesh metal screen or slats of wood. The lining may be installed some distance from the surface of the insulation so as to protect the stored fish from heat entering at the sides and bulkheads. Cooling grids can be installed in the air space, protected by the lining. A lining over the insulation at the deckhead is often unnecessary. The floor insulation is commonly covered with concrete.

VAPOUR SEAL

In building a cold store great care should be taken to ensure that the vapour seal is complete and efficient. Since there will be much less water vapour in the air in the cold store than in the air outside (the vapour pressure in the cold store will be lower than outside) there will be a tendency for moisture to travel through the insulation. If the water vapour is allowed to pass into the insulation, it will condense and freeze in the insulation itself, reducing its value and in severe cases physically destroying it.

Quite a few materials have vapour sealing properties, for example metal foil, some kinds of plastic film and steel plate. Their suitability may depend to a great extent on the particular application. An efficient vapour seal can be made by trowelling two coats of bitumen on to a reasonably smooth inside surface, on the warm side of the insulation, before the insulation is installed. The vapour seal must envelop the store completely, without a break. Points where pipes, conduits, etc., pierce the insulation should be sealed.

Some insulation materials, such as expanded ebonite and plastic, have good vapour sealing properties. Even so, there usually will be need for a separate vapour seal. Particular care should be taken to seal any joints, especially where there might be movement on warming up and cooling down. Compounds which remain plastic at low temperature and folded membranes have been used for this. When the insulation is installed in the form of slabs, special attention should be paid to joints where the slabs are butted together. The insulation should be in more than one layer with the joints staggered so that there are no joints straight through the insulation.

INSULATION

Fishroom insulation has been discussed in Chapter 4. The insulation in the cold store serves the same purpose as the insulation in the wet fishroom but, of course, the store temperature is much lower.

The heat gain through the insulation, usually by far the major part of the cooling load for the store, will be reduced by increased thickness of insulation. There comes a point, however, where increased insulation thickness is not justified. Not only may the cost of added insulation fail to bring a worthwhile reduction in the required refrigeration capacity but also the volume of the cold store will be reduced. In a store of about 750 m^3 the design heat gain through the insulation typically will be reduced from 16 to 11 kJ/s by an increase of 2·5 cm in insulation thickness that would reduce the size of the cold store by about 20 m^3. Too little insulation, on the other hand, not only implies an unacceptably high cooling load but also may promote dehydration and deterioration of the fish. It is recommended that the average thermal resistance of the areas bounding the cold store should be not less than $2{\cdot}5\ m^2$ degC/W. This usually implies an insulation thickness of more than 10 cm overall, and at least 4 cm over the ends of the frames. A somewhat greater amount of insulation may be necessary in the warmer climates. The design temperature difference between the outside temperature and the cold store temperature normally will be in the range 60 to 80 degC, allowing for solar radiation.

There should be sufficient insulation covering structural members. Ideally the insulation should be unbroken but, where it is necessary, a material of relatively high thermal resistance such as plastic or wood often can be used in place of metal for connections to structural members, conduits, pipe hangers, etc.

A large number of ships have been fitted with plastic (polyurethane) insulation foamed in place. It has the advantage of good insulation value and is impervious to moisture with no joints. Awkward shapes present little difficulty with this type of insulation, although care must be taken to ensure complete coverage and reasonably uniform density. It can be applied by means of spraying or, if the space to be insulated is suitably bounded by steel, wood, or other material, the insulation components can be mixed and poured into the space. It can be poured between lining and hull.

Loose and blanket insulations by themselves are not used often for the cold store. They may be inclined to settle, depending on the method of fixing, and have no vapour sealing qualities. Their thermal resistance is not particularly high so that, for a given insulation value, the thickness will have to be greater than the thickness for most other types. They are used sometimes in combination with other types because of their cheapness, for example for insulation between the frames, where the insulation over the ends of the frames is of the board or foamed type.

Most of the vessels for freezing at sea are constructed of steel. Expanded insulation materials which are impervious to water are preferred, to some extent because they are virtually unaffected by any leaks in the hull. Repairs to the hull and insulation can be carried out easily.

Fire risk is an important consideration in any insulation system but especially on board ship. Insulation linings should be fire resistant. Special precautions are necessary where plastic insulation materials are used, during construction and repair work as well as during normal operation because of the toxic fumes which are produced when they burn.

COOLING SYSTEMS

Cooling systems for the wet fishroom are discussed in Chapter 5. Those of interest for the cold store are the pipe grid with natural convection of air, forced convection cooling and the jacketed fishroom.

Pipe grids

Grids of plain galvanized steel pipe, which are widely used, are generally considered to be the best type of cooler for the cold store. In a well designed store, grids are installed at the deckhead and down the sides. Thus much of the heat is intercepted by the grids before it can reach the fish. Since there should be an air space all around the store anyway, grids at the side, preferably shielded by an open type of lining or slats, need not imply a loss of valuable space. Enough space should be left, however, for natural convection of air around the grids. The space should be 10 cm or more all round though it may be less where the amount of pipe grid installed is generous. Stored material should not come into contact with the grids. With this system the rate of air circulation is modest throughout the store and there are no severe local temperature differences which may lead to dehydration.

The amount of grid to be installed depends on the design cooling load and the temperature difference between cold store and refrigerant. Some allowance should be made for the accumulation of frost on the grid. As frost is deposited on the grid initially, the resistance to heat transfer is increased slightly but it soon reaches a more or less constant value and further accumulation of frost has no effect until it begins to interfere with natural convection of air. The grid size should be based on a value of overall thermal resistance of not less than $0{\cdot}12\ m^2\ degC/W$. The temperature difference between cold store and refrigerant should be low, not more than 10 degC and preferably 5 degC. Thus for a difference of 5 degC the cooling surface required is at least $2{\cdot}4\ dm^2/W$.

The pipe grid can be operated for long periods without a defrost. In vessels with ample grid area, defrosting is not carried out during the voyage and perhaps for several months. The methods employed are brushing of the grid and allowing the store to warm up when it is empty.

Finned pipe grids sometimes are installed in order to reduce the first cost and save space. Care must be exercised in choosing the amount of grid to be installed because the reduction in cooling capacity due to frost formation on the grid will be greater than with plain pipes. In an extreme case where the fins are close together, the gap between fins will be bridged by frost very quickly and the grid made even less effective than if there were no fins at all, giving a great reduction in cooling capacity. The fin spacing should not be less than 2·5 cm. Periodic defrosting, usually about once per week, will be necessary. Since brushing of the grids is awkward and can be difficult or impossible when fish has been stowed, a hot defrost should be included, but this means that dislodged frost and water from deckhead grids will fall on to the cargo.

Forced circulation of air

Forced circulation of air by means of fans with a compact cooler as for the air blast freezer, Chapter 8, has been preferred for many cold stores ashore because of reduced cost of construction and ease of defrost. Defrosting usually is carried out electrically or by hot gas.

Because of reduced cost and saving in space, forced circulation systems have been installed in some vessels but they have not given as good results as plain grids. The fan and cooler may be installed as a unit in the store proper or the air may be circulated and distributed through ducts more or less along the lines of air conditioning systems. As in the blast freezer, it is imperative that the fan heat, a potential source of dehydration, is removed by the cooler. Distribution of air should be uniform and velocities everywhere in the store should be low. Local strong currents of air will encourage dehydration. Wrappers and containers for the fish may provide little protection. There can be considerable damage due to movement of moisture within the pack itself.

As in the grid system, the fish should not be stowed directly against the insulation. The heat entering the store through the insulation should be intercepted by an air space before it reaches the fish.

Semi-jacketed systems, similar to those for wet fishrooms described in Chapter 5, have been devised. The air can be circulated between linings and insulation. A system of ducts can be installed to provide cooling all around the store. Normally there is ample room between the hull frames for both insulation and ducts; ducts can be formed from the insulation and linings by themselves.

More data is needed on the performance of the better systems with forced air circulation, as opposed to gravity circulation with grids.

The jacketed store

In the jacket system, discussed in Chapter 5, a sealed air space is included between the insulation and the inner lining of the store as shown in Fig 5.1. It is effective in minimising dehydration; the store humidity remains high. Heat gains through the insulation are absorbed in the jacket and so it may be feasible to use a lesser thickness of insulation. It will be necessary to install a small cooler in the cold store proper, not shown in the figure, in order to cope with other cooling loads imposed by items such as air changes and lights. There need not be a loss of space with this system because produce can be stowed directly against the inner lining. There is little experience and cost data available for this kind of store for fishing vessels, but in most cases it will not be worthwhile compared with the ordinary method of insulation with grids.

Chapter

Freezing at Sea 10

APPLICATION

FREEZING at sea enables vessels to remain longer on the fishing grounds and also land fish of high quality. Increased fishing time is an important consideration in distant water fishing where, with ordinary icing techniques, a high proportion of time is spent travelling between port and fishing grounds. Initially vessels equipped for freezing the catch at sea were introduced widely in order to cope with the increased demand for fish, especially demersal species, of good quality in the face of decreasing rates of catch on grounds remote from the home port. Following extensions of fisheries limits in the 1970's the scope for such operations has declined somewhat, at least in some countries. Nevertheless often freezing at sea will be preferred to chilling, on grounds of fish quality, since chilled fish will be of prime quality for only a few days, or in some cases hours, after catching.

With facilities for freezing instead of chilling the catch the criteria on which design is based are changed. Less emphasis is needed on speed of transport and handling and more on storage capacity. On board the freezing vessel, therefore, less space may be allotted to propulsion machinery and more to the cold store. Generally the vessel is larger with a larger crew, particularly when the fish is filleted or processed in some other way before freezing on board, as opposed to the freezing of whole fish. Refrigeration and processing machinery must be accommodated. Since the duration of voyage is relatively long, special attention must be paid to the facilities for the well-being of the crew, recreational, education, medical, etc., as well as good working conditions. Recruitment of suitable crew has become difficult where the standard of living and availability of work ashore have improved. There will be an increasing emphasis, therefore, on automatic methods of catching, navigation, processing and freezing in order to reduce the demands on the crew.

Freezing of the first part of the catch combined with storage in ice for the latter part has been practised on only a few vessels. It has the disadvantages that the refrigeration plant must be more or less the same size as in the vessel which freezes all of the catch, since the required freezer output will be much the same, and the building and operation of the ship becomes more complicated.

The advent of freezing at sea has been accompanied by changes such as the change from side trawling to stern trawling which, apart from advantages of improved safety, easier handling of the fishing gear, etc., lends itself to good factory layout. Instead of gutting fish on deck in the open, the work can be done between decks, out of the weather. Improvements in fishing gear and in fish machinery for handling, gutting, filleting, etc., will make further reductions in the very heavy labour of fishing. In the arctic fishery, however, increasing comfort in the factory has meant higher temperatures which can affect the fish adversely. In any event particular attention must be paid to chilling of the fish and control of conditions before freezing. In warmer climates the need for control is more obvious.

In fact freezing on board is being used more widely in waters relatively near to port. Fish from catcher vessels have been transferred to factory or freezer vessels in port or in sheltered waters. The freezing and cold storage plant need not be on a fishing vessel at all but can be on a converted cargo vessel or a barge with or without its own propulsion. Normally in such cases it will be feasible to include a fish meal plant. This approach is attractive in seasonal pelagic fisheries because it can enable the processing of large catches of highly perishable fish. Mobility gives important advantages since the delays and losses associated with the transport of chilled fish will be markedly reduced.

In this chapter, the freezing of fish at sea will be considered mainly from two points of view; the freezing of whole fish and the freezing of fish products, principally fillets. Some of the problems encountered are common to both methods.

Processing before freezing

The amount and sort of processing to be carried out before freezing varies from one fishery to another and from one country to another. It is largely a matter of choice. The fish may be frozen whole without any treatment beforehand or they may be cut into portions, fillets, etc., before freezing so that there is a minimum of treatment ashore before the fish is consumed. Frozen tuna, herring, cod and other species are supplied to processing factories ashore where they may be canned, filleted, smoked and so on. On the other hand, these processes can be carried out on board factory ships which may depend on their own catches or partly or wholly on catches from other vessels.

Many factors are involved in making the choice between the factory ship with a high degree of processing and the vessel which simply freezes the catch with little or no processing beforehand. The first cost and the operating cost of a factory vessel are usually relatively high without a corresponding high catching rate, hence the interest in fleet fishing and transfer of the catch at sea from one vessel to another. Typically, the factory will have a capacity for 40 t of raw material per day but often the average catching rate has been less than 20 t per day.

A general improvement in catching methods, giving an increase in catching rate, has helped to make the factory ship more attractive. Automatic devices and better machines for processing the fish also have played an important role, permitting a wide range of operations and easing the work of the crew. There has been some development of machines for sorting according to size and species and orientation of the fish. Improved machines for weighing at sea have been developed but usually measurement is by volume, as in the production of fillet blocks frozen in trays in the horizontal plate freezer and most freezing methods. Machines capable of separating bones from flesh also are available. They produce a fish 'mince' and enable the recovery of material which would be discarded otherwise, or converted to fish meal, for example flesh from fish skeletons from the filleting machine.

The development of machines and processes for underutilized resources such as blue whiting, Antarctic krill and small squid is necessary before they can be fully exploited.

There has been hardly any freezing of consumer packs of fillets at sea. Fillet freezing has been confined to large blocks for catering establishments, restaurants and the like, and for the production of fish portions and fish sticks ashore. Often the fillets are interleaved with paper or plastic film in order to keep the cut surfaces clean and prevent them from sticking together on thawing, or make it possible to separate the fillets while they are still frozen.

Factory ship

Large factory ships which process the catch to the point where it is more or less ready for the consumer can remain away from the home port for long periods, typically for three months or more. Normally much of the raw material supplied to the factory is converted to fish meal, widely used for the feeding of farm animals. The high degree of processing on board means that the stowage rate of material for human consumption is much higher than the stowage rate with whole fish. Although this has obvious advantages, there is no doubt that the recruitment of crew for the longer voyages is a problem at least in some countries. Under these circumstances the crew may be changed from time to time in order to limit the period of duty away from home.

Sometimes the factory ship will act as a mother ship, dealing with the catches from ancillary vessels. A flotilla consisting of mother ship and catcher vessels may be employed or the mother ship may carry the catcher vessels on deck. The catcher vessels may employ refrigerated sea water or ice supplied by a mother ship equipped with a flake ice machine. The factory ship may be a converted vessel such as a tanker or passenger ship. The fish may be supplied directly to the home market or to foreign markets by the factory ship, or it may be transferred to another vessel for shipment.

Fleet fishing

Transfer of the catch at sea from one vessel to another can take place in various ways; by derrick, elevator, pump, floating container, bag or codend, etc. In some cases the catch is stored for a time on the catcher vessel before transfer, and in other cases transfer takes place directly after catching. The transfer of iced fish in boxes has been mentioned in Chapter 2.

There are difficulties, however, associated with transfer of the catch. Fishing must be arranged so that loss of fishing time due to transfer is held to a minimum. There can be serious spoilage of the fish, depending on the conditions during transfer. Even ½ h in the sea in a codend, for example, may give rise to appreciable losses of quality and mechanical damage to the fish by virtue of movement of the mass. These faults may be reinforced by delays before processing on board the factory ship.

The freezing of whole fish

The freezing of whole fish can be carried out on board the factory vessel, of course, and because of its relative simplicity, it can be employed on fishing vessels of various types such as trawlers, tuna longliners and seine net vessels. The main advantages of freezing whole fish compared with the method of processing on the factory ship are the lower first cost for the vessel and often a lower operating cost. The size of crew is smaller, typically less than half the size for the factory vessel without a corresponding difference in the yearly catch, although the larger vessels will be able to fish in rougher weather and spend more time on the fishing grounds.

The freezing of whole fish means that there is a good deal of latitude in the way it can be processed ashore because it can be virtually equivalent to fresh chilled fish. In many respects processing ashore is cheaper than processing on board. On the other hand, the cold store space required will be much greater for whole fish and it often will be necessary to thaw the fish on an industrial scale before processing and distribution. Processing may involve freezing for a second time but this is acceptable under good processing and handling conditions where a small change in texture can be tolerated.

Various methods of thawing are employed; heating in humid airblast and in circulating water or water sprays and in steam under vacuum at about 18°C are common methods. In some cases it may be feasible to combine thawing and brining, for example the thawing and brining of frozen herring for canning by immersion in heated brine. Thawing has been employed on the factory vessel where whole fish have been frozen and stored during periods of heavy fishing and then fed to the processing line during periods of slack fishing.

Stowage rate

The stowage rate of frozen fish will depend on a number of factors and there can be large differences depending on the type of freezer, the type of packaging and method of stowage. Some typical values are given in Fig 10.1, as a rough guide. Allowance has been made for the fishroom structure and for the packaging of the large blocks of fillets and the consumer packs. The density of the block of whole fish has been taken as 800 kg/m^3, based on the blocks from the vertical plate freezer.

It can be seen that the stowage rate on the factory ship is much higher, hence the longer voyages. With a hold capacity of 600 t and a catching rate of 12 t of whole (gutted) fish per day, about average for the arctic fishery, the vessel freezing whole fish will remain on the fishing grounds for 50 days. Sometimes heading, removal of the heads of the fish as discussed in Chapter 2, is practised in whole fish freezing in order to increase the stowage rate of edible material and so prolong the voyage. It has not been practised on the majority of vessels, however, because it increases the work of the crew, there is the problem of loss of fillet material which could be largely solved by better machines for heading, and the heads are of some value to

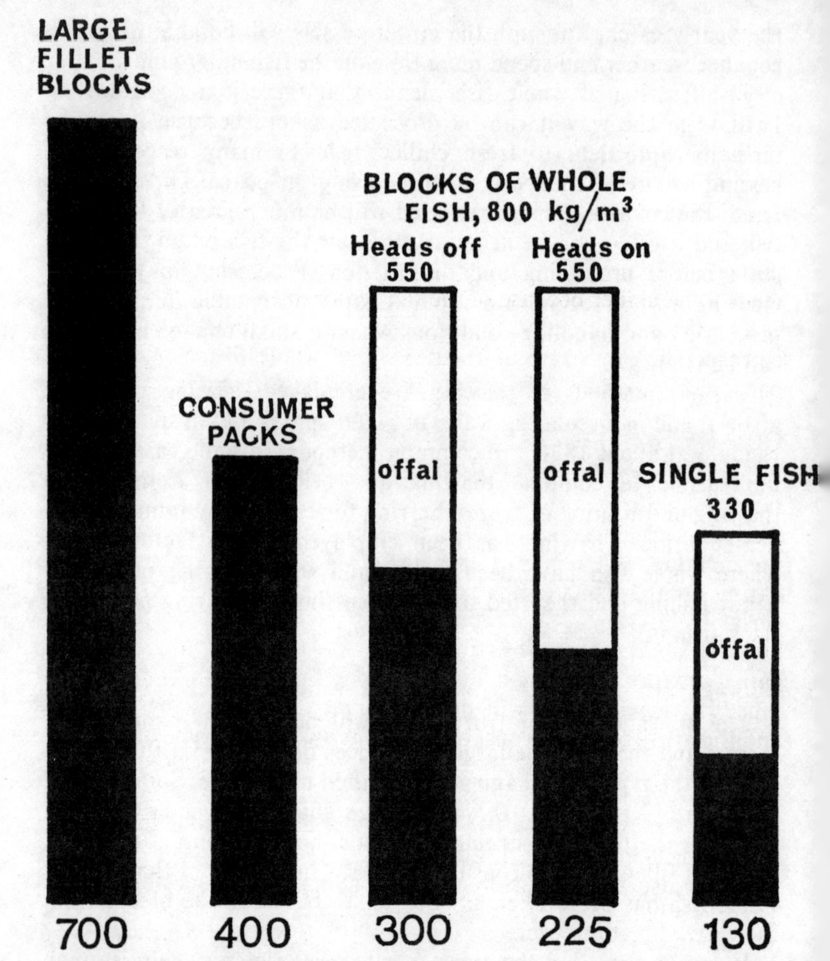

FIG 10.1 Typical stowage rates for frozen cod

factories ashore, notably fish meal factories. The problem of obtaining crews for the longer voyages also has been a factor.

Refrigeration capacity

The freezing capacity required depends on the fishery and the amount of processing on board. Vessels with a crew of up to thirty for the freezing of whole fish usually have a freezing capacity of between 1½ and 2 t/h. This can cope with all but the heaviest rates of catching in arctic waters where the average output is less than 1 t/h. The power required by the refrigeration plant is much less than that required for propulsion of the vessel. The total electrical load imposed by a Refrigerant 22 or ammonia plant operated at −45°C suction temperature, with a rated output of 1½ t/h, is roughly 300 kW, allowing for freezers, defrost, cold stores and pump circulation. The power for systems with air blast freezers tends to be higher because of the fan requirements.

The typical factory ship will have a capacity of 1¼ t/h of fillets, corresponding to a rate of about 3 t/h of whole fish. Some vessels, particularly mother ships, have a much higher capacity.

Quality Factors

Although good results can be obtained by freezing, it is important to realize that there can be difficulties in the freezing of very fresh fish which do not arise when chilling is used as the method of preservation. In the main these difficulties are not associated with eating quality but with appearance and handling properties; shrinkage and breakup of the muscle, associated with *rigor mortis*, possibly accompanied by some loss of fluid and toughening of eating texture, and blood discolorations in the flesh. The importance of these factors will depend largely on processing methods and the attitude of the consumer, in other words the type of market. The fish may be of excellent eating quality even though it may suffer from defects of discoloration and breakup but in some markets, for example for fish fillets where appearance is important, these defects are serious disadvantages.

There also will be differences depending on the species of fish. There is a paucity of published observations, especially on full scale operations as opposed to experiments. Much of the following information in this chapter is based on observations on the freezing of whole fish of various kinds and on the freezing of fillets from fish such as cod, redfish and hake. It will give a general idea of the problems encountered in freezing very fresh fish, a subject on which more experience is needed.

Rigor mortis

It is convenient to consider the interval between catching and freezing at sea in terms of *rigor mortis*, death stiffening. Shortly after death there begins a period of time during which the flesh tends to become hard and the muscles contract due to chemical changes. The time required to complete the changes depend on the species, temperature and physiological condition of the fish. In a fish that has been exhausted by struggling after catching or in one that is in poor nutritional condition, much of the energy reserves which can cause contraction will be absent. The forces of contraction can be quite strong and can create problems in processing and freezing. They can be of significance in some species such as cod but the reactions in others, hake for example, have relatively little effect. Typically, the time required for cod to pass through *rigor* is 3 days at 0 °C, 20 h at 10 °C and 10 h at 17 °C. The intensity and effects of *rigor* are at a minimum just above the freezing point and increased by higher temperature. The contraction forces can be very strong at 17 °C.

A fillet taken from a fish before *rigor* sets in can shrink a great deal, accompanied by a toughening of the flesh and loss of fluid. Fillets which have been cut from cod before *rigor* have a dull, matt surface which does not produce a good smoke cure.

In the whole fish the skeleton restrains these contractions, preventing shrinkage, but the contraction forces can cause breaking up of the flesh, particularly at high temperature. In cod at 17 °C or higher, there will be considerable damage, partly due to a weakness at high temperature of the tissues which connect the muscle blocks. Rough handling of the fish such as the forcible straightening of a bent fish in *rigor*, even at low temperature, also can result in damage and is to be avoided at all stages on board. This is an important point, especially for freezing at sea, since freezing itself weakens the connective tissues to some extent and will slightly increase any tendency toward separation of the muscle blocks.

The chemical reactions responsible for *rigor mortis* will proceed very slowly in fish which has been frozen before *rigor* and stored at −30 °C. If the duration of storage at low temperature is short and thawing is sufficiently rapid, the reaction will cause *thaw rigor*, with the normal tendency for contraction of the muscle. This can occur in cod after more than 18 months at −30 °C. In practice it creates no difficulties with whole fish but can cause shrinkage, etc., in small portions such as individual fillets and fish sticks during thawing or when thawing and cooking are combined. Slow thawing or a period of high temperature storage, say one or two days at −5 °C, permits the chemical reactions to proceed while the fish is frozen stiff and contractions are avoided. Deliberate treatment of this kind must be carefully supervised in order to prevent spoilage.

Bleeding

Discoloration of the flesh of the fish can be a serious disadvantage, more so in some markets than in others, although it may have no effect on eating quality. Discoloration can arise in various ways; through bad gutting which will permit some deterioration due to enzymic action, through bruising of the flesh during handling and through insufficient bleeding. Bruises, which are blood marks in the flesh, are caused by the rupture of blood vessels due to rough handling. It is obvious, therefore, that rough treatment should be avoided.

The flesh of fish which has been properly gutted and stored in melting ice without an excessive delay after catching will bleed under ideal conditions. The result is that blood discoloration is not a major problem where the chilling of gutted fish is the method of preservation. If the flesh is frozen before there has been adequate bleeding, however, the thawed flesh will be stained with blood. These stains in the thawed fish cannot be removed by washing.

There are wide variations in important physiological characteristics which determine the ease of bleeding. As a general rule, the temperature of the fish and surroundings should be low, below 5 °C. Thus the blood will not congeal rapidly in the flesh before the fish is cut and clotting is prevented so that the blood can escape after the fish is cut. If the cut is made while the fish is alive, the heart will pump out most of the blood and the time required for adequate bleeding will be short. On the other hand, the longer the dead fish is held before bleeding, depending on the temperature, the more difficult and prolonged bleeding will be.

The cutting of the fish for bleeding can be done in various ways. The throat or tail may be cut or the head may be removed. These methods and also gutting give more or less equal results under normal conditions. Hence gutting of the fish usually will be sufficient for the purpose of bleeding. Where emphasis is placed on early bleeding of the catch, sometimes the cutting for bleeding is done as a separate operation as soon as possible after catching, even when the fish are to be gutted later on. After gutting, or cutting, the fish should be left for a time to bleed at chill conditions, preferably in ice or in chilled water, possibly for $\frac{1}{2}$ h or more depending on the species and other factors.

The best method known of bleeding cod, redfish and some other species has been possible only experimentally so far. The head is removed and the tail cut. Then the main blood vessel which runs next to the backbone is flushed out with cold water, giving very quick results if the fish are bled soon after catching.

If the fish are to be frozen as fillets, even closer attention will have to be paid to bleeding if blood discolorations are to be avoided. The interval for bleeding may have to be longer than for whole fish freezing. Blood present in the flesh tends to come to the surface of the fillet after it has been cut. This can be a problem, especially with fillets cut before the onset of *rigor mortis*, where the surface blood cannot be removed by washing. In any event, fillets should not be placed in contact with water because it may cause them to shrink and there may be some leaching and waterlogging of the flesh.

Gutting

Gutting and bleeding are not always carried out. Herring, for example, is usually frozen whole, without any treatment beforehand. Tuna are not always gutted. The freezing of whole, ungutted cod has not been considered worthwhile because the presence of the guts can lead to the development of undesirable flavours during storage and interfere with thawing and processing. Gutting is considered to be a prerequisite to filleting in many applications.

Improvements in the gutting operation have brought an improvement in bleeding through quicker and gentler handling. Gutting benches and conveyor systems have replaced the traditional method of gutting in pounds in many cases and gutting machines are in use on some vessels. Gutting has been discussed in Chapter 2.

Delay before freezing

If the fish are to be held for a time before freezing, chilling in ice or in water, usually sea water at about 0°C, should be employed. The permissible length of delay at chilled conditions prior to freezing varies according to conditions. Observations on the eating quality of fish held in cold store have shown that the storage life depends on the length of time in chilled storage before freezing. In many cases appearance factors also will restrict the permissible delay before freezing. In a typical case the storage life of white fish fillets at −30°C will be 30 weeks where the gutted whole fish have been held in ice for 9 days prior to freezing and 100 weeks where the fish have been held in ice for 1 day.

For gutted cod which is to be cold stored at −30°C for up to one year, thawed and then processed, perhaps with freezing for a second time, the limit in ice should not be more than 3 days. The corresponding limits for hake, the flesh of which is softer than cod, and haddock will be lower.

Ungutted herring should not be held in ice for more than 24 h before freezing and in cases when they are particularly susceptible to self-digestion because they have been feeding, the limit is less than 24 h. Off flavours will develop in herring during cold storage if the period in ice is too long.

The limits in refrigerated sea water will be lower, partly due to the penetration of salt into the fish which causes undesirable changes during cold storage. Experience has shown that the limit for hake and cod in refrigerated sea water should be set at about 12 h.

Quality Control

In the world markets more and more emphasis is being placed on high quality and uniform quality. Whereas with iced fish a reasonably good estimate of quality can be made on landing, immediately before distribution, the quality of frozen fish is not assessed easily until it has been thawed. Two aspects of this question are the need for good methods of quality assessment of frozen fish and a good understanding of the relationships between the condition of the fish immediately before freezing and the final quality. Some of the difficulties that arise due to variations in the intrinsic condition of the fish caused by factors such as starvation, feeding, spawning and parasites are known.

It is possible to exercise a good deal of control on quality on board. A complete log of operations, including data on the refrigeration plant should be kept. This can be linked to a system of labelling the frozen fish according to species, size, fishing grounds, treatment on board and possibly an assessment of intrinsic condition. It is good policy to label fish that has suffered undue delay before freezing or been given a non-standard treatment, perhaps due to heavy catches or breakdown of plant, so that it can be identified after landing. Operations from the landing of the fish on deck to the stowage in cold store should be carried out with an appreciation of the refrigeration and quality aspects.

Very often it will be in the interest of the catcher to see that the procedures ashore, transfer, cold storage and thawing, are properly carried out.

Systems for Freezing at Sea

Practical systems for the freezing of whole fish and fillets on board trawlers have been developed since about 1950. Early attempts did not meet with much success because the techniques were poor and there was a lack of facilities for proper handling, storage and thawing ashore.

A common procedure with cod, shown in Fig 10.2(a), is to gut the fish as soon as possible after they are landed on deck, then wash them in more or less the same way as in wet fishing, described in Chapter 2, before freezing. In fact the entire operation is the same as for icing, except that the washing time may be extended to several minutes in order to extend the time for bleeding. A disadvantage is that no refrigeration is applied before the fish is put into the freezer, except in some cases where the temperature of the sea water in the washer happens to be lower than the fish temperature. There is little delay between stages except during heavy fishing which overloads the facilities on board and results in delays before gutting of more than 4 h, often at high temperature. There can be delays between gutting and freezing if the gutting rate exceeds the freezer output. This sometimes occurs during heavy fishing with hand gutting. Sometimes, with small fish, the gutting cannot keep pace with the freezers. Delays before freezing also have occurred where part freezer loads have been held back until enough material was available to make up a full load. Batch freezers should not be too large and they should be operated in sequence, as discussed in Chapter 8.

The practice on some vessels of slitting the throats as a separate step as soon as possible after catching, shown in Fig 10.2(c), gives better results because more time is allowed for bleeding and, since gutting by hand is a slower operation, bleeding will be carried out earlier and hence more effectively. A separate washer may not be necessary when a gutting machine is used, as a washing stage consisting of jets of water is included in gutting machines.

When the fish are filleted, usually by machine, the filleting is done as a last step before freezing. The interval between filleting and freezing should be as short as possible. It is feasible to arrange the filleting and freezing so that the maximum delay never exceeds 1 h.

These methods of handling are similar to the normal practices associated with icing but the fish may suffer damage due to high temperature before freezing. Also, there is insufficient time for bleeding between gutting and freezing, where gutting only is employed for bleeding, and conditions for bleeding are far from ideal, particularly when deck and washer temperatures exceed 5 °C. This has led to the installation of facilities for chilling the catch before freezing and more controlled conditions for bleeding. Clearly

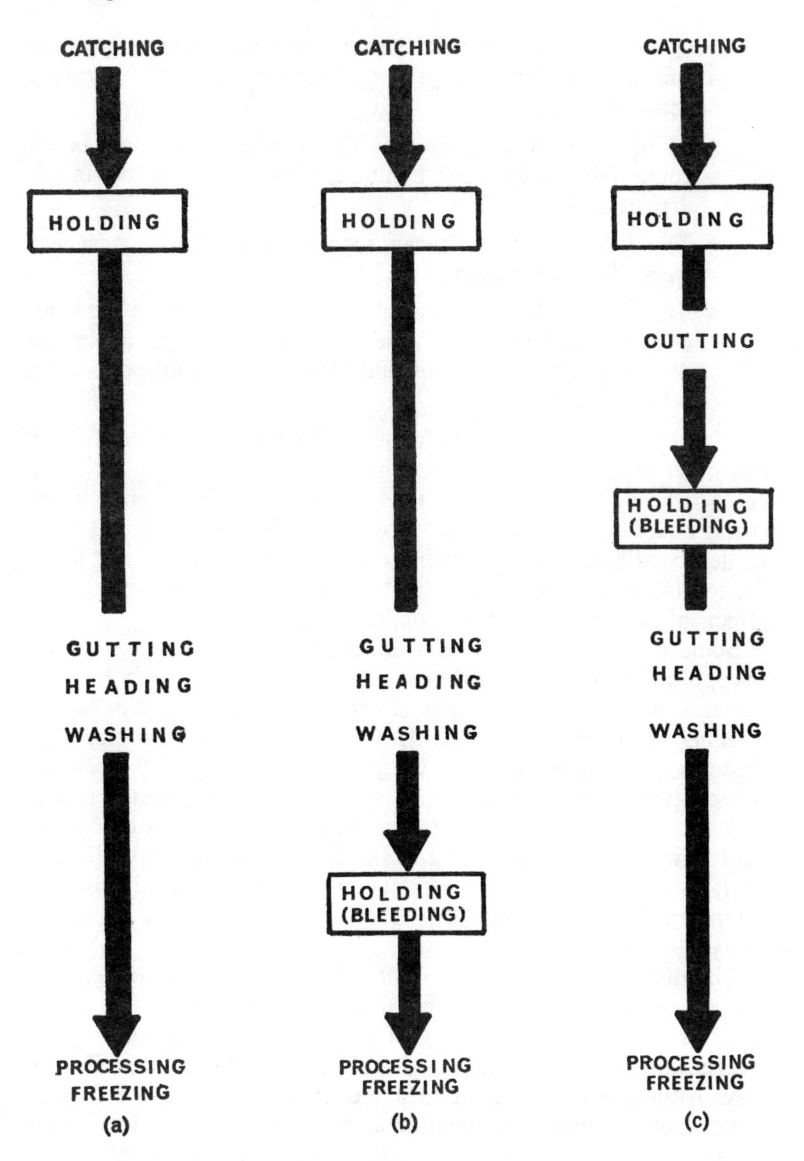

FIG 10.2 Handling before freezing

from the earlier sections of this chapter, the major features required of a system for handling fish prior to freezing are low temperature and punctual handling.

- As in wet fishing, outlined in Chapter 2, there should be no contamination or rough handling of the fish from the time it arrives on board.
- The fish should be frozen soon after catching, consistent with adequate bleeding.
- The fish should be held at chill conditions, below 5°C, from catching to freezing in order to retard spoilage, avoid the ill-effects of *rigor mortis* and make the bleeding operation more effective.

Where whole fish are to be frozen without gutting or bleeding or any other treatment beforehand, the question of handling after the catch is brought aboard is relatively straightforward. Some chilled storage may be necessary before freezing, as any delay should be at chilled conditions. At the other extreme, where the fish are to be filleted before freezing close attention must be paid to the possible effects of *rigor mortis* and the conditions for bleeding.

Some of the newer systems for freezing at sea include chilled storage before freezing. With the procedure shown in Fig 10.2(c), bleeding is carried out in tanks of refrigerated sea water or in ice. Chilling is applied after the fish have been cut for bleeding. Systems also have been developed for holding in tanks of refrigerated sea water before and after gutting, as indicated in Fig 10.2(b). They have the important advantage that the catch can be chilled immediately on coming to the deck. It is relatively difficult to apply refrigeration in an effective way to a heap of fish on the ramp or in the gutting pounds, either by means of ice or sea water sprays.

The difficulties created by *rigor* contractions in cod in filleting operations can be reduced or eliminated by holding the fish in ice for 2 or 3 days before filleting. The variations in demand on the freezers due to uneven catching rate also can be reduced in this way. Although it can result in a reduction in the size of freezer installation required, the practice of holding the fish in ice before freezing usually has been avoided because of the problems of providing space and labour for the operation. It may be the best approach, however, where a catcher vessel holds fish for later transfer to another vessel for freezing.

Passage through *rigor* can be accelerated by storage at elevated temperatures. Proposals have been put forward in which a compromise is made; the fish would be bled at low temperature and then stored for a short period at an elevated temperature of, say, 10 °C. The period of storage might be reduced to about 24 h in this way with acceptable quality and no *rigor* effects. Since icing does not give the desired temperature and refrigerated sea water should not be employed for such a long period it would be difficult to effect the desired degree of control. Again extra space and possibly extra labour would be involved. Thus there would appear to be too many disadvantages to the method.

Chilling

The required amount of refrigeration can be estimated with the help of the information in Chapter 1. If the initial fish temperature is 20 °C, for example, the refrigeration requirement for the fish only will be 80 MJ/t, corresponding to a fish to ice ratio of about 4 to 1. Probably about 3 to 1 would be used.

Ice can be supplied by a flake ice machine on board or it can be carried in the cold store and supplied as required. Ice has an advantage over mechanical refrigeration in that ice can be supplied as required whereas the mechanical system will be overtaxed during periods of heavy fishing unless there is ample capacity. Ice can be used in a refrigerated sea water system with or without supplementary mechanical refrigeration, as described in Chapter 6. Nevertheless, because of the ease of handling the catch and simplicity of application, refrigerated sea water tanks with mechanical refrigeration will be preferred in most installations.

With good design and operation, the total energy required for refrigeration will not be increased by a chilling system. A mechanical chilling system operating at a suction temperature of −5 °C, say, will extract heat in the chilling range more efficiently than the freezer system operating at a temperature of −45 °C. This is illustrated by Fig 7.1. A chilling plant also can be used for air conditioning on board.

REFRIGERATED SEA WATER SYSTEM

The refrigerated sea water, RSW, system described in Chapter 6 has formed the basis of chilling systems for improved handling on freezing at sea vessels, the major differences being that the permissible period in RSW will be much shorter and there will be some changing and replenishment of water in order to maintain clean conditions. According to the procedure outlined in Fig 10.2(b), the RSW system would be divided into two separate parts, before gutting and after gutting. They could employ the one refrigeration plant. A reserve of water should be included in both, so that only a limited amount of refrigerated water is wasted at a controlled rate, bearing in mind that there should be some store of refrigeration for the rapid chilling of the larger catches. In this way the amount of refrigeration for the chilling of incoming water can be limited to less than 20 per cent of the total for chilling. RSW temperature would be maintained below 5 °C and as near as possible to 0 °C. All tanks should be completely insulated with at least 2 cm of insulation giving a thermal resistance greater than 0·5 m^2 degC/W, otherwise the cooling load will be higher than necessary. The insulation must have complete protection against the ingress of moisture. Impervious insulation is preferred, particularly where an exposed vapour seal may become damaged.

According to one method outlined by Fig 10.2(c), the first holding period is very short and without chilling. The fish are cut and then held under chilled conditions until required. This method makes demands on manpower and space but is practicable where emphasis is placed on bleeding before filleting on board.

Batch systems, along the lines of the example below, are at an early stage of development. They give some chilled storage capacity for heavy fishing conditions but normally with a limiting period of 12 h or less. Continuous plant eventually may be more attractive in some installations, fitting better into a continuous production line and requiring less supervision from the crew.

RSW before gutting

A number of tanks instead of only one or two will enable segregation of catches, effective chilling, orderly processing and periodic cleaning. The storage before gutting should be in three or more tanks with facilities for draining or pumping out the water and emptying. In some cases it will be convenient to build the tank with a sloping bottom so that the fish are more easily removed for gutting. It may be necessary to rake out some of the fish. In one system, a perforated platform has been installed at the bottom of the tank. The platform is hinged at one end and can be raised at the other end in order to tip the fish on to a conveyor which removes them from the tank. Provision must be made for charging the tank with fish at the top.

The washing stage of gutting machines can be supplied with waste (refrigerated) water from the system.

The reservoir can be equipped with a refrigerated pipe coil or an external heat exchanger can be used, as described in Chapter 6.

The refrigeration capacity will be based on the amount of fish to be chilled when there is heavy fishing, the requirement for make up water and heat gains through the tank insulation, pump heat, etc. Pumping rates should be 4 to 6 dm^3/s for each 10 t of fish to be chilled. Separate pumps for each tank are preferred to a single pump with a manifold system.

A typical specification for a vessel capable of freezing $1\frac{1}{2}$ t/h of whole fish, or the corresponding amount of fillet material, is as follows:

storage capacity, maximum fish to water ratio 4 to 1	4×5 t (fish)
reservoir capacity	10 m^3 (water)
pump capacity	4×3 dm^3/s
make up water	$\frac{1}{2}$ t/h
approximate refrigeration capacity of heat exchanger	140 kJ/s

The value for refrigeration has been based on ambient, water and fish temperatures of 20 °C and the cooling of a catch of 20 t of fish over a period of 4 h. It includes allowances for the make up water and for heat gains and pump heat. Somewhat lower values might be used, depending on the fishery. The refrigeration capacity required on trawlers in the arctic fishery will be a little more than half the above amount.

The interval between charging the tank and draining it should not be too long but neither should it be too short for effective bleeding later on. So that a sufficient amount of cooling is applied, an interval of not less than $\frac{1}{2}$ h is recommended.

RSW after gutting

Storage or 'bleeding' tanks after gutting should be designed for ease of loading and unloading. Tanks with steeply sloping sides have been satisfactory. There should be several small tanks with separate pumps and a common reservoir and heat exchanger. Cod and hake would be held for at least $\frac{1}{2}$ h when they are to be frozen as whole fish and at least 1 h when they are to be frozen as fillets, to be filleted immediately before freezing. In the typical specification below, again based on the conditions given above, the refrigeration capacity for the fish will be less than half the total cooling requirement, since more than half the chilling will be done before gutting.

storage capacity, maximum fish to water ratio 4 to 1	6×1 t (fish)
reservoir capacity	4 m^3 (water)

pump capacity	$6 \times 0{\cdot}5\ dm^3/s$
make up water	$\frac{1}{4}$ t/h
approximate refrigeration capacity of heat exchanger	70 kJ/s

Total cooling load

The capacity of the compressors need not include a full allowance for both totals, the capacities of both heat exchangers, because the two systems will not be under maximum load at the same time and the reservoirs will have some smoothing effect. Thus a compressor capacity of 160 kJ/s, with the RSW at 0°C, would be sufficient in this case and possibly only 110 kJ/s under arctic conditions.

TUNA

The freezing of tuna warrants special mention because of the importance of the fishery and the problems peculiar to the freezing of large whole fish such as tuna. There are considerable differences in fat content and other characteristics, spoilage pattern and spoilage rate depending on the species of tuna. They vary in size from 2·5 to 60 kg or more and their initial temperature can reach as high as 32°C in tropical waters. The average size of fish in the catch of yellowfin tuna has been about 10 kg and more than 20 per cent have been over 12 kg which corresponds to a thickness of about 20 cm. The normal maximum thickness is 30 cm, for a fish of 60 kg.

One method of fishing for tuna, purse-seining, yields large catches which impose problems in handling. Although the average catching rates have been less than 10 t/day, most of the fish is caught in periods of heavy fishing when a single set of the seine net produces more than 10 t and occasionally more than 100 t. This is in contrast to the longlining method, where the catching rate is more even and not more than 8 t/day during approximately 10 h of line hauling.

Although the spoilage rate of tuna is lower than that of most other fish, the shelf life of chilled tuna may be 30 days whereas it is 14 days for chilled cod, the normal rules for the freezing of whole fish should be applied if high quality is to be maintained. Gutting, washing and bleeding should be carried out soon after catching and chilling should be applied during any delays before freezing. Large whole fish should be frozen individually in order to achieve maximum rates of freezing and for ease of handling. The plate freezer is unsuitable. Basically two systems are used for whole tuna;

freezing in sodium chloride brine and air blast freezing. These methods are described in Chapter 8. Sometimes low air velocities are employed and the fish are placed directly in contact with the coolers, formed by pipe coils or plates which may act as shelves. The air blast method is similar to that for other large fish and presents few quality problems but the method of freezing in brine is a special case.

As mentioned in Chapter 1, some whole tuna is frozen in air blast and stored at temperatures substantially below −30 °C, sometimes as low as −50 °C, in order to prevent colour changes in the frozen flesh and thus meet special demands for raw tuna.

Freezing in brine

Much of the whole tuna frozen at sea is for later canning ashore where the presence of some extra salt in the frozen material can be tolerated. Large quantities for canning are frozen by immersion in brine and sometimes by brine spray, mostly on purse-seine vessels with a fish capacity of 400 to 900 t. This is the largest application of the method on fishing vessels.

The ideal procedure from the point of view of quality is to freeze the fish soon after catching and then deposit them in a cold store. Where it is necessary to chill the fish as a separate step, the period of chilled storage should not exceed 5 days. A freezing capacity of $1\frac{1}{2}$ t/h will be sufficient for most purse-seine vessels freezing their own catches. A disadvantage of freezing in brine is that the temperature of the fish cannot be reduced much below −15 °C but, for long periods of storage, the fish should be held at a temperature of −20 °C at most and preferably −30 °C or below after freezing.

There are a number of methods employed. Freezing by immersion in brine or in a brine spray may be followed by cold storage under dry conditions. Salt penetration can be held to a low level. A system of freezing the tuna in wells, insulated watertight compartments, is employed on a larger number of vessels.

According to one method used on purse-seine vessels making voyages of 6 to 12 weeks, the fish are placed in the wells for chilling, freezing, cold storage and finally thawing on board. The wells vary in size. A typical well 3 m by 3 m by 7 m deep, with a holding capacity of 50 t of tuna, will have roughly 50 m^2 of cooling surface in the form of plain steel pipe grids at the sides of the well. The grids are refrigerated by a direct expansion ammonia system and in some cases are supplemented by an external heat exchanger for cooling the brine. The fish are not gutted. They pass directly into the wells where they are held in circulating refrigerated sea water for usually not more than 8 days until the well is full of fish, 800 kg/m^3, and the temperature has been reduced to near 0 °C. Salt then is added to the sea water, normally to make up a 15 per cent solution with a freezing point of about −11 °C, and the temperature is reduced, usually by adjustment of a back pressure regulator to lower the evaporation temperature. The cold brine is circulated through the mass of fish in order to freeze them. After a freezing period of 24 to 72 h the brine is pumped out of the well and the grids are used directly to maintain the desired storage temperature. The fish must be at least partially thawed in order to remove them from the well on landing. This is done by circulating warm brine through the well. Unloading of the fish in this way results in increased salt penetration and sometimes physical damage to the fish.

Quality for canning can be acceptable under ideal conditions with the system of freezing and storage in refrigerated wells. In practice, however, it is variable and often low because of the delay before freezing, high fish temperature during freezing and cold storage and poor thawing conditions. Salt penetration tends to be high. Freezing capacity usually is substantially less than 1½ t/h and the density of the fish in the well sometimes exceeds 800 kg/m^3, interfering with the flow of brine. A typical circulation rate of brine for the well of 50 t capacity is 30 dm^3/s using a 7½ kW pump. The cooling capacity of the grid usually imposes a limitation on the rate of freezing, particularly with the smaller fish. The grid of 50 m^2, operated at about −20 °C during freezing, will have a capacity of not much more than 30 kJ/s, corresponding to a freezing rate of less than ½ t/h. Brine temperature is in the region of −5 to −7 °C. The grid temperature during cold storage is −7 °C or lower. The system has the advantages of minimum handling at sea, very high storage rate and low power consumption for refrigeration but is not suitable for fish other than tuna. Suggested improvements include increased freezing capacity, the installation of external heat exchangers where they are not already in use to supplement the grids, reduction in the temperatures of freezing and storage, increased brine velocity and the use of removable containers instead of wells for easier discharge. It is doubtful, however, whether the method could be adapted for the production of fish of high quality for a variety of uses.

The routine with the brine spray method is much the same. The brine spray is installed at the top and recirculated from the bottom of the well or hold where the pump intake is located. In some vessels the brine spray system has been fitted into tanks of the type used for refrigerated sea water, described in Chapter 6, giving a more versatile vessel. An external heat exchanger is used and no pipe grids are installed in the tanks. The tanks usually have been insulated for this purpose and in some vessels jacket cooling has been used in the same way as in the jacketed fishroom described in Chapter 5.

SHELLFISH

Freezing at sea has been employed in fisheries for shellfish such as shrimp, lobster and crab. By and large the principles involved for crustaceans and molluscs are the same as those for other fish. Air blast, plate and immersion freezers have been used for boiled and unboiled material, depending on the product and the application. Heads and some other parts of the fish may be discarded in order to increase stowage rate. The freezing of only the meat can result in large reductions in weight and volume. With some shrimp, for example, the reduction in weight will be in the ratio 1 to 3 and in volume 1 to 6.

Often it is advantageous to freeze shellfish individually. Although the stowage rate may be lower with individually frozen fish, there can be advantages in distribution, sale and thawing, bearing in mind that shellfish is relatively high in value. Plate and air blast freezers have been used for shellfish and meat from shellfish in baskets, trays and packs. Air blast and immersion freezers have been used for whole fish, headed fish and sections in the shell such as tails, claws and legs. They are attractive where the shapes are irregular and where a quantity placed in a basket of wire mesh, bag or similar receptacle takes up a loose, open form, perhaps with a bulk density of less than 300 kg/m^3. Thus cold fluid can circulate through the batch, tending to give rapid freezing. Immersion freezing has been employed to some extent for shrimp and other crustaceans. Salt brine has been used extensively but it does entail some salt penetration into the meat which is increased by cutting or severing the fish into sections. Good results in freezing shrimp have been obtained with solutions of salt, sugar and water. Shrimp frozen in an agitated solution of corn syrup and salt water at $-20\,^{\circ}C$ tend to remain separate, individually frozen, and the freezing time is only a few minutes. The fish are glazed by the solution and there is no need for glazing as a separate step. Whole and headed shrimp have been frozen in this way in baskets and in mesh bags which are used for storage and distribution. Plant especially designed for the freezing of shrimp by immersion in a solution of corn syrup and salt water, incorporating compact and reliable refrigeration units and suitable cooling for the fish hold, has been installed in a number of vessels. These solutions and ordinary saturated brine, with a salt concentration of about 26 per cent, can be cooled readily to the freezing point of approximately −21 °C and therefore accumulate ice during periods when there is little or no fish to be frozen. In this way the refrigeration capacity during periods of high cooling load can be increased.

In some fisheries it is standard practice to cook the shellfish, notably lobsters and shrimp, in boiling sea water or salt water before freezing. The reasons are not always clear and sometimes it is undesirable. The cold storage life will be reduced, depending on the amount of salt present, although glazing and packing in vapour tight packs and vacuum packs will increase the storage life. There are cases, however, where the removal of the meat from the shell of species such as lobster and crabs is easier if they are cooked before freezing. Often the meat is removed by hand, perhaps with the aid of jets of compressed air or sea water jets. Some peeling machines are available, but, as with other kinds of fish, there is scope for the development of improved machines for the processing of shellfish, for sorting and grading and for the separation of meat and shell.

Index

Other books published by Fishing News Books Limited Farnham, Surrey, England

Free catalogue available on request

Mending of fishing nets
Modern deep sea trawling gear
Modern fishing gear of the world 1
Modern fishing gear of the world 2
Modern fishing gear of the world 3
Modern inshore fishing gear
More Scottish fishing craft and their work
Multilingual dictionary of fish and fish products
Navigation primer for fishermen
Netting materials for fishing gear
Pair trawling and pair seining—the technology of two boat fishing
Pelagic and semi-pelagic trawling gear
Planning of aquaculture development—an introductory guide
Power transmission and automation for ships and submersibles
Salmon and trout farming in Norway
Salmon fisheries of Scotland
Seafood fishing for amateur and professional
Ships' gear 66
Sonar in fisheries: a forward look
Stability and trim of fishing vessels
Testing the freshness of frozen fish
Textbook of fish culture; breeding and cultivation of fish
The fertile sea
The fish resources of the ocean
The fishing cadet's handbook
The lemon sole
The marketing of shellfish
The seine net: its origin, evolution and use
The stern trawler
The stocks of whales
Training fishermen at sea
Trawlermen's handbook
Tuna: distribution and migration
Underwater observation using sonar